AFV 2914

24.15 / 18.71 1

WITHDRAWN

The Loggerhead Turtle

in the Eastern Gulf of Mexico

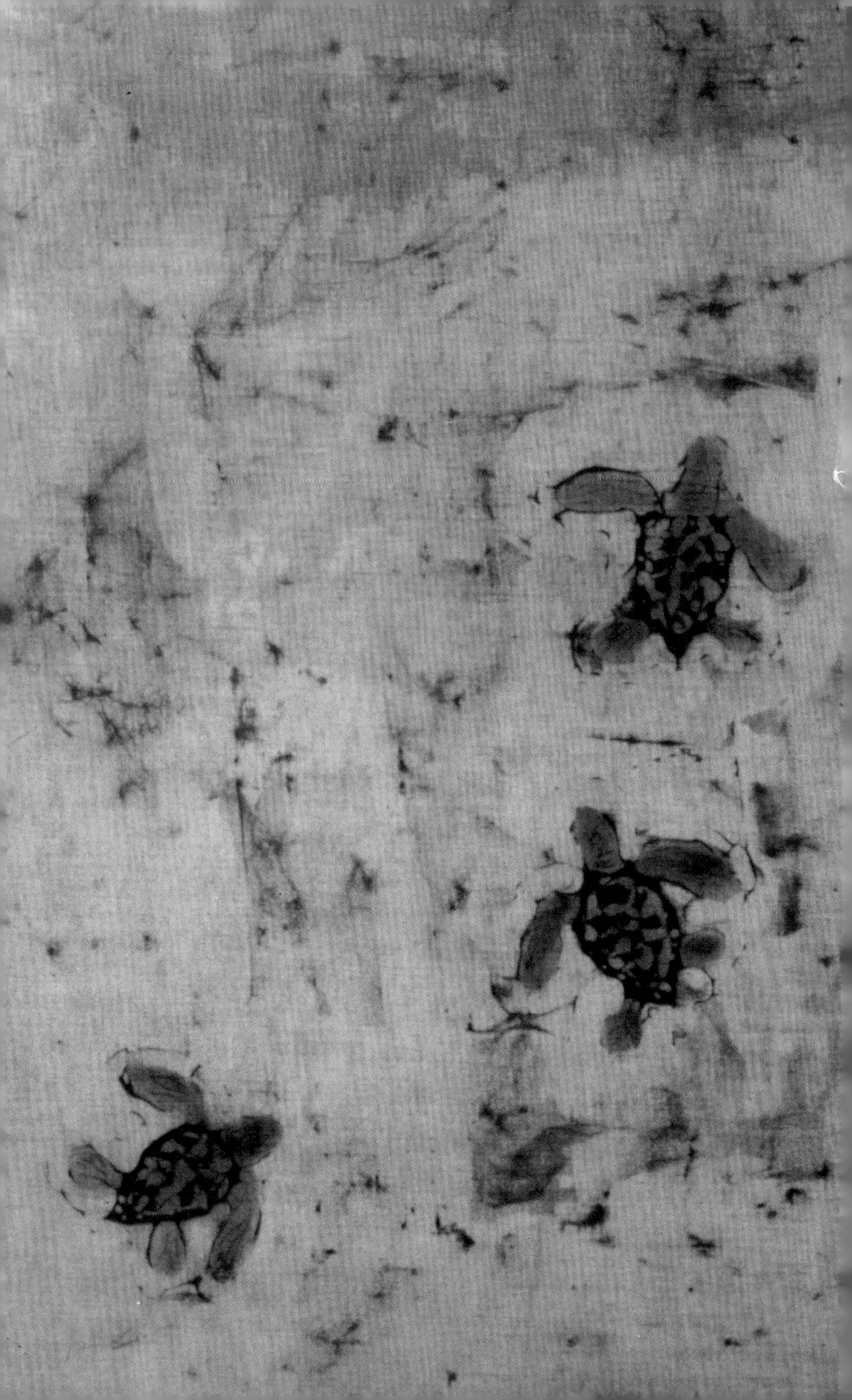

The Loggerhead Turtle

in the Eastern Gulf of Mexico

Charles R. LeBuff, Jr.

Caretta Research, Inc.
Sanibel, Florida

Copyright © 1990 by Charles R. LeBuff, Jr.

All rights reserved. No part of this book may be reproduced or utilized in any form or by any means, electronic or mechanical, including photocopying, recording or by any information storage and retrieval systems, without permission in writing from the author.

Printed in the United States of America

Library of Congress Cataloging in Publication Data

LeBuff, Charles R., Jr., 1936-
The Loggerhead Turtle in the Eastern Gulf of Mexico
Bibliography: p.
Includes Index
I. Turtles. II. Sea Turtles. III. Loggerhead Turtle. IV. Title.

Library of Congress Catalog Card Number 89-81763
ISBN 0-9625013-0-1

Published by:

Caretta Research, Inc.

With major funding provided by:

Adelaide Cherbonnier
Jane Widener Forsyth
The Damroth Foundation
The J.N. "Ding" Darling Foundation
The John E. and Aliese Price Foundation
The Sanibel-Captiva Conservation Foundation

For My Parents—

Who had the courage and foresight to sever roots and relocate their family to Southwest Florida . . .
before it was raped!

Contents

Part 3

Preface

Turtles have held a special fascination for me for more than forty-five years. As a small child I would wander the New England woods and stalk pond margins searching for eastern painted, spotted, and northern snapping turtles. For hours on end each summer, I studied the occupants of a turtle exhibit at a local zoological park, carefully noting the behavior and characteristics of each species. Never in my wildest boyhood dreams, however, did I envision myself as an adult on a darkened white sand oceanic beach, conducting similar observations on gigantic counterparts of those turtles of my childhood.

As a youngster, while most of my cohorts were reading comic books, I was delving into such fascinating reading as Harold Babcock's *The Turtles of New England* and Clifford Pope's *Turtles of the United States and Canada.* In 1952, the publication of the classic work of Dr. Archie Carr, *Handbook of Turtles*, increased my level of interest in the group of marine turtles. His later books about sea turtles, *The Windward Road* and *So Excellent a Fishe*, generated continuity of this interest into my adult life.

In late June of 1953, I had the opportunity to see my first loggerhead turtle crawl and nest on Bonita Beach in southern Lee County on Florida's lower Gulf coast. A year later, I saved my first loggerhead turtle and her nest just a few miles to the south on Vanderbilt Beach in Collier County. A group of high school classmates decided to kill a loggerhead turtle which had innocently blundered into a summer night's beach party as she searched for a nesting site. I was able to muster enough support among my peers and we insisted that the hapless turtle not be harmed—and she wasn't. Moments later, I kneeled there in the moonlight and watched in awe as a timeless prehistoric scene of nest construction and egg laying unfolded before me. I have since observed this primeval act of procreation countless times, and each time I continue to be deeply moved. I had discovered a special role for myself—one could perhaps claim it as a niche controlled by destiny—for on that very night my life's work was created.

Word soon spread through the small town of Naples of my concern for turtles. Whenever residents discovered a turtle crawl on their beach they notified me and, each summer thereafter for several years, I doubt if a loggerhead turtle made a landing on Naples Beach that I didn't hear about. I would examine each crawl to determine if it were a nest or not, and then mark every nest with a coded stake. I soon learned the period required for egg incubation, and many times I'd be present to witness the remarkable sight of enthusiastic hatchlings emerging from the sand, scurrying across the beach, and safely reaching the surf.

In 1958, I relocated to Sanibel Island to accept a position on what is now the J.N. "Ding" Darling National Wildlife Refuge, a 5,200 acre unit that is administered by the U.S. Fish and Wildlife Service. In May of 1959, following the urging of W.D. "Tommy" Wood and J.N. "Ding" Darling, I began what was to become, as far as I know, the oldest uninterrupted loggerhead turtle monitoring program in the United States. Today, this work is ongoing and continues to serve a vital role in conservation and field research relative to the life history, ecology, and survival of loggerhead turtles.

"Tommy" Wood was the first manager of the refuge and "Ding" Darling was a syndicated editorial cartoonist who escaped Iowa winters to spend them on Captiva Island. He was, among other things, twice a Pulitzer Prize-winning cartoonist, once Director of the Biological Survey, forerunner of the Fish and Wildlife Service, and designer of the first Federal "Duck Stamp." Mr. Darling was also an advocate of preserving Sanibel and Captiva Islands, and he vigorously pushed for the establishment of the Sanibel National Wildlife Refuge. The Refuge was renamed in his honor in 1967, a few years after his death.

In 1968, because of increasing community awareness, the monitoring program became known as Caretta Research with a board of directors composed of individuals who had a genuine interest in the conservation of sea turtles. The goals of the group were developed and closely integrated with the original loggerhead turtle population monitoring facets of my work. The following objectives guided this program for two decades:

1. On-going research into population trends within the marine turtle community of Southwest Florida.

2. Maintaining surveillance of sea turtle nesting assemblages along the Southwest Florida coastline by both aerial and on-the-ground surveys.

3. Expanding the organization's sea turtle tagging program in Florida.

4. Studying factors of nesting and production success at a complex of study sites.

5. Relocating loggerhead turtle eggs to increase nesting colony productivity, and to mitigate destruction of eggs because of predators and adverse environmental impacts.

6. Maintaining a limited number of young loggerhead turtles in captivity to study growth patterns, evaluate tags and other methods of identification, and pursue educational goals.

7. Developing a program of public education publicizing the plight of all marine turtles; sponsoring marine science workshops; and developing audio-visual packets for distribution to schools, libraries and service organizations.

8. Cooperating and exchanging information with other marine turtle conservation groups in the United States.

For five years, Caretta Research was an autonomous organization associated with the Sanibel-Captiva Conservation Foundation, Inc. (SCCF). Contributions from interested persons passed from the SCCF to Caretta Research as grants which enabled us to fund our specialized conservation programs.

In 1973, two major changes occurred. Caretta Research became an incorporated, tax exempt, not for profit foundation and separated from SCCF. The geographical area of our concern was also enlarged. The tax exempt status qualified us for foundation grants and larger membership contributions, each necessary to carry on the various activities. Most beaches and communities along the Southwest Florida Gulf coast, between Clearwater and Cape Sable, were included in the monitoring and tagging programs with a representative from each beachfront community serving as a Unit Leader. By 1977, fourteen Unit Leaders conducted the field programs of Caretta Research, Inc., in just as many communities along the lower Gulf coast.

The field studies which were implemented at these Units served many purposes. Foremost, in their initial start-up phase, was the population survey and tagging program. Later, some subregional loggerhead turtle nesting populations were found to be severely impacted by predators. Nesting success at those Unit study beaches was enhanced by adoption of successful egg relocation and protection programs. Increasing the scope of the field operations substantially expanded the volume of biological data being collected and further enhanced the loggerhead turtle studies which had originated on Sanibel Island.

In 1969, I wrote a report, *The Marine Turtles of Sanibel and Captiva Islands, Florida,* that was distributed to the membership of SCCF. This booklet, now long out of print, was a progress report on my early island sea turtle work and it included some of the basic data produced. It was an elemental publication, one that established long-term goals and identified needs. Its publication kindled a personal desire to continue the field work and later to produce a comprehensive work on the biology and life history of the loggerhead turtle. The challenge of the past twenty years has been recorded within these pages.

Early on, while assembling the first draft of this manuscript, I decided that the finished product should reach the layman—the wildlife and conservation oriented public—and that its subject matter should not be directed solely to the scientific community. This book is a semi-technical treatise on the biology, life history, and everyday problems which confront the population of loggerhead turtles occupying the eastern Gulf of Mexico. Therefore, I have chosen to indicate measurements, distances, and temperatures in United States customary units. Occasionally, because of standard measuring techniques, i.e., egg sizes and shell dimensions, I have utilized a combination of the above and metric units. This has not been done to confuse the reader, but rather to give precision to the measurements. For comprehension of the various measurement standards and definitions of scientific terms used, I have furnished a glossary on page 203.

In the very early days, I measured carapace lengths and widths over the curve of the shell, and it wasn't until much later when sea turtle studies gained popularity elsewhere, that straightline dimensions became the standard format. I have continued to collect both curvature and straightline measurements. When shell measurements are recorded, the length is taken from the nuchal indent of the anterior carapace back to the most posterior marginal extension. Width measurements are collected between the widest points of the carapace. These areas of the loggerhead turtle shell are identified in Figure 2 on page 9.

Throughout the text, I have used a combination of common and scientific names to specifically identify plants and animals. Scientific names that are used have been adopted from general reference works on the various forms of flora and fauna and are current nomenclature. Since biological taxonomy is not infallible, over time some of these names will change. Occasionally, common names for marine animals are not available; they simply do not exist, i.e., some benthic organisms. They have been omitted for that reason.

Sanibel Island
January, 1990

Acknowledgements

Because of the passage of time, it is difficult to recall the many wonderful and caring people who have contributed time, talent, and money in support of my sea turtle conservation pursuits. If the following lines omit anyone I should have included, please accept my apology and very special thank you.

My wife, Jean, was always tolerant and supportive of my nighttime activities chasing loggerheads up and down the Sanibel beach. For over thirty-two years she has encouraged me to maintain my interest in sea turtles. Every so often, as I labored over this manuscript, Jean would peer over my shoulder, read the copy on the computer screen, and whisper something that I had overlooked or suggest some important commentary to my work. My children, Leslie and Chuck, were my constant companions during their young years, and great helpers when I otherwise would have had to make my nightly beach patrols alone. Later, as a young adult, Chuck assisted field teams at many of my study beaches.

The program would never have continued without the sheer dedication and invaluable assistance provided by very special coworkers such as Eve M. Haverfield, Richard W. Beatty, Edward J. Phillips, Paul W. Zajicek, Patrick D. Hagan and James P. Anholt. For many years, Eve handled such tasks as fund-raising, membership, hatchery projects, and volunteer services for Caretta Research, Inc. I appreciate her commitment. Richard, Ed, Paul, Pat and Jim were always ready to pitch in within their areas of expertise and do their share, making every effort that special study projects were completed or that motorized equipment kept on rolling. For several years in the 1970's, Paul was my chief assistant and supervised the work of Caretta Research, Inc. on the beaches of Cape Romano, in Collier County. Several of Pat's excellent photographs are an integral part of this book.

The technical drawings and some of the other original artwork, unless a credit line specifically indicates otherwise, were created by my lifelong friend Warren E. Boutchia. Warren is an outstanding wildlife artist and professional medical photographer. I would not have launched this writing project without knowing he would be my key illustrator. As children, over forty-five years ago, we talked of writing a book about turtles someday.

The frontispiece illustration is that of an original batik made by an unknown artist . . . It depicts three loggerhead hatchlings heading for the surf and beyond, on a perilous journey to meet their destiny.

Several professional photographers contributed their best sea turtle material to be included in this volume. The name of each appears beneath this respective work as a credit line. Where no credit line is present the photograph is my own. Graphics and other identified illustrations were rendered by the talented Mary Lou Schadt.

Illustrations on pages 19, 20, 21, 24, 184, 190, 194, and 198 are published through the courtesy of Dr. L.D. Brongersma and the Rijksmuseum van Natuurlijke Historie, Leiden, The Netherlands.

In addition to the above, the following people were actively engaged in a variety of responsible positions over the years or otherwise rendered invaluable service to the conservation of loggerhead turtles in Southwest Florida and to the objectives of Caretta Research, Inc.: Ardis Allen, Dolores Ambrose, Alice Anders, Armand and Beverly Ball, Rick Bantz, Tootsie Barnes, Roy Bazire, Kathy Boone, Chris Brown, George Campbell, Judy Carberry, Glenn Carowan, Helga Cernicek, Ralph and Billye Curtis, Sue Dapcic, Roger Exline, Jane Fitzhugh, Bruce Frazier, Anina Glaize, Gloria Gram, Bill Hammond, Phyllis Harned, Diana Harris, Karla Heimann, Betty Hoffman, Bill Ihle, Laurie Ihle, Dan James, Helen Jans, Jimmie and Jim Jones, Joan Kain, Doug Kenefick, Caroline Legette, Erick Lindblad, Mike Lubich, Paul Manley, Jenny Mapes, Bob Marek, David Marsoli, Mary McHarg, Carol Miller, Wayne Parker, Tucker Patton, Bob and Lucy Pond, Mike and Dorothy Saunders, Laura Schuchard, Debbie Shoss, Kristie Seaman, Carol Sellers, Lanny Sherwin, Ann Silvis, Edythe Stokes, Lynn Stone, Ty Symroski, Dick Thompson, Leslee Tucek, Catherine Turner, Jim Vanas, Sarita Van Vleck, George and Mike Weymouth, Ann Winterbotham, Esperanza Woodring, Mo Woolverton, Fran Wright, and George, Betty, and David Zajicek.

Other than the individuals and organizations listed on the title page, the following members of Caretta Research, Inc. made special financial contributions to support the production and printing of this book: Bob Averill, Dr. and Mrs. K.C. Emerson, Charles S. Estabrook, Jr., Mr. and Mrs. Edwin H. Ewing, Anina Hills Glaize, Mr. and Mrs. Porter J. Goss, Dr. and Mrs. Ned N. Kuehn, Mariner Properties, Inc., Mr. and Mrs. Frederick Rudolph, Dr. Howard and Brenda Sheridan, Sarita Van Vleck, and Ann L. Winterbotham.

I also appreciate the long-term cooperation provided by the Florida Department of Natural Resources, in particular Ed Joyce, former Director of the Division of Marine Resources, and his staff. When the permitting requirements for sea turtle conservationists were being revamped back in the 1970's, Ed assigned Turtle Permit Number 001 to me. There are now nearly two hundred such permits issued annually by the Department. I would also like to acknowledge the sea turtle enforcement efforts made over the years by the many Officers of the Florida Marine Patrol who were assigned to Southwest Florida.

Dolores Ambrose, Laura Schuchard, Edythe Stokes and Leslie Young typed early draft sections of the growing manuscript.

Dr. Joan Girard graciously provided the statistical analysis of data dealing with egg sizes and egg shell thicknesses.

Dr. Lew Ehrhart, of the University of Central Florida, and Paul Zajicek each reviewed early drafts of the manuscript and made important recommendations.

Martha Ambrose made an editorial review of the final manuscript and rendered many helpful suggestions.

CHARLES LEBUFF

Introduction

Charles LeBuff and I, both of whom love turtles, are often asked why. The standard reply I have evolved is simply: "All children are interested in turtles. Some of us simply never grew out of it." On the other hand, a different selective process operates when one works with sea turtles. Everyone is happy to spend a few pleasant, insect-free hours on a turtle beach, watching one or two turtles nesting before retiring to a comfortable bed for the night. But to patrol a beach all night, every night, come hellish mosquitoes or high water, is definitely a process that separates the men from the boys. So there is a paradoxical yet admirable, child-man quality to those, such as Charles LeBuff, who patrol a turtle beach every season for three decades in a row. The obsession with turtles is immoderate, perhaps, and incomprehensible to those who make their living selling real estate, or who live on submarines, or who work in factories. Yet there is envy too; the pure happiness of the turtle field man, doing exactly what he most enjoys, having the thrill of the hunt without the massacre of the innocent that climaxes a traditional hunt, and the contact with nature, weather, and the fixed, primordial cycles of tide and moon, are rarities in today's frenetic, synthetic world.

I like and admire Charles, not just because he is loyal to his turtles through thick and thin, but also because he is loyal to his friends. He has a host of them, and I felt distinctly honored to be the one selected to write this Introduction. Dozens, if not hundreds, have served in the uniformed ranks (yes, they even have dress uniforms) of Caretta Research, and, whatever their ultimate station in life, they will never forget their service on Sanibel, Captiva, Keewaydin, and the other beaches of the lower Gulf Coast. And it is fully in character that many of the illustrations for the book were done by Warren Boutchia, whom Charles has known since his childhood nearly fifty years ago. Charles was even a friend of "Ding" Darling, in his grave a quarter century now, and known to the modern generation of conservationists as an icon of the past, the patronym of the J.N. "Ding" Darling National Wildlife Refuge, rather than as an actual human being.

I learned a lot from Charles' book. There is a lot to be learned when one stays in one place so long, keeps one's eyes open, works with a small population of turtles, and studies them so intensively that they have few secrets remaining. I was intrigued by Charles' invention of the "turtle stopping device," a sort of giant hairpin that one pokes into the sand in front of the turtle, so that its shoulders are restrained and it can be held in one place for tagging and measuring. I do the same thing by kneeling in front of a turtle, with a knee thrust between the turtle's head and foreflipper on each side. But although this has worked well for me (I have a long upper leg), I have recommended it to others who have tried it with painful and potentially disastrous results.

I learned too of the extraordinary productivity of the loggerhead turtle. Multiple nesting by this species has been recorded many times in many places. But to have witnessed each of the six nestings by a single individual turtle in a season, and get an exact egg count each time, can only be done on a thoroughly patrolled beach with low density nesting. Charles' recording of his turtle CR 140, which produced 920 eggs in the course of six nestings in 1973 is a unique and remarkable observation.

Just as Charles stays with his Sanibel beaches year after year, so do most of his turtles. But some are wanderers, and his records of individual turtles nesting on Sanibel one time and later on Melbourne Beach, well up on the Atlantic coast, the next time; or of the one that nested on Jupiter Island one time, and then later on Sanibel, are unique examples of non-fixity of nest site in sea turtles.

Charles too has developed strong views, and much information, on the menace posed to sea turtles by shrimp trawling efforts. This is the hottest and most serious issue in sea turtle conservation today. The rumors that Charles reports, of the wheelhouses of trawlers festooned with strings of turtle tags as decorations, emphasize the difficult nature of the problem.

But best of all I liked the account of the honeymooning couple being disturbed by a nesting loggerhead during a romantic tryst on Upper Captiva Island. This event is not without precedent. Indeed, such an occurrence has a place in the permanent folklore of the Palau Islands, in the Western Pacific. According to this legend, a young couple repaired to a remote beach on Ngemelis Island, and after making love, used the girl's grass skirt as a pillow to sleep. When they awoke in the morning, the grass skirt had disappeared, and there was a fresh turtle nest right beside them. Embarrassed at what seemed like discovery while they were asleep, they nevertheless returned to the same beach fifteen days later, and just as they were about to sleep, a turtle emerged from the ocean with the remains of a grass skirt still attached to its flipper. They knew from this that it was the same turtle, and ever since, the people of Palau know that, if they find a fresh turtle nest, they may be able to catch the turtle itself if they go back in fifteen days.

The warm Gulf breezes blow through Charles' account of a thousand starlit nights, of ten thousand sandy miles, patrolling the shell-strewn shores of Sanibel and Captiva in search of the silent giant reptile that finally emerges, a blacker hulk against the blackness of the sand, phosphorescent breakers shattering over the humped carapace, a silhouette at once familiar and mysterious; of countless hours crouching in the sand beside the same dark form performing its ancient, deliberate rituals of digging, of laying, of covering; a universe of one man and one reptile beneath the stars, two creatures similar in size yet separated by hundreds of millions of years of divergent evolution—but now, by only a few feet of distance; the human brain contemplating the infinity of the cosmos and the tiny reptilian brain programmed and automatic, without thought for the work it commands; yet both playing out their destiny side by side in the sand, a crossing of paths of two life forms, one of the land and one of the sea.

Charles loves his islands and his turtles, and we can share that love when we read his book.

PETER C.H. PRITCHARD

December 1989

Part 1

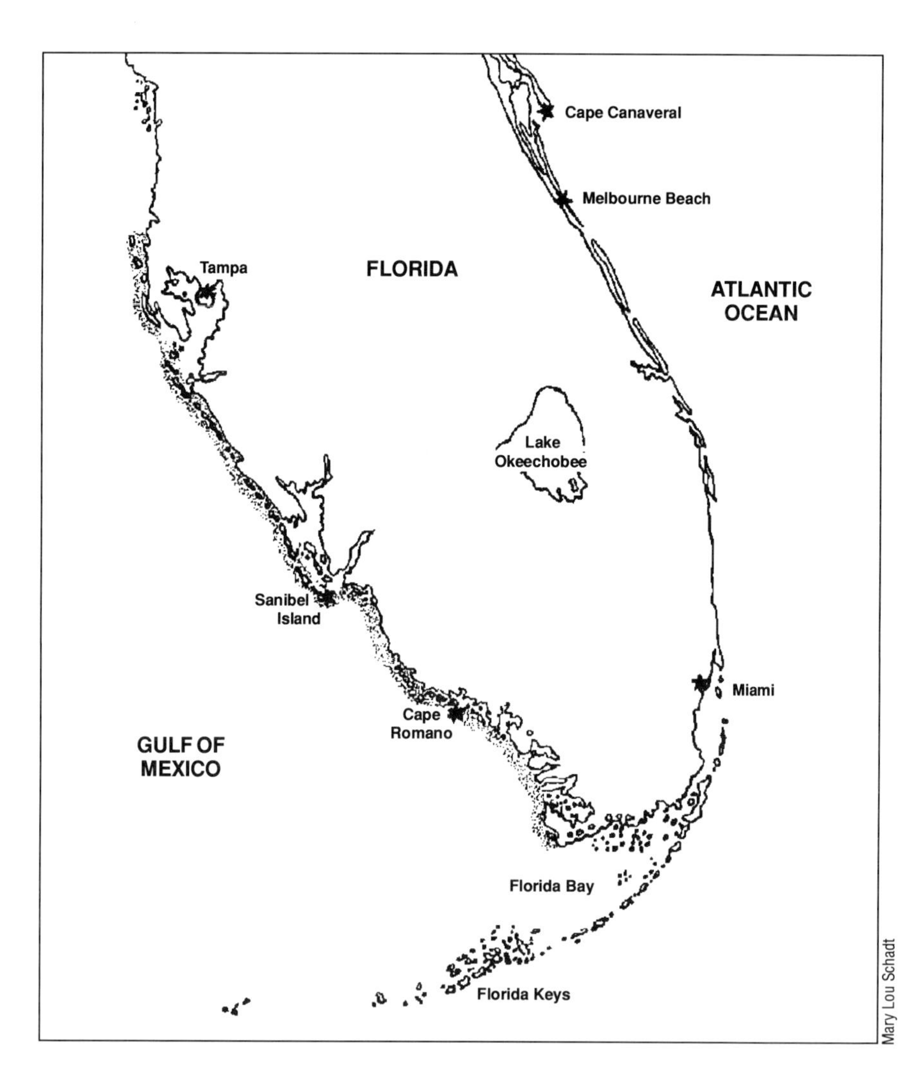

The Subregion

1
The Subregion

The majority of the western Florida coastline is located on the shores of the eastern Gulf of Mexico. This region plays an essential role in many aspects of sea turtle ecology. Most of the field work, the results and interpretations which are detailed herein, were accomplished along the Florida Gulf Coast. Geographically, this subregion can be identified as waters and lands to the east of the one hundred fathom line, or those marine systems above 24 degrees north latitude and east of 86 degrees west longitude. I have designated the Southwest Florida Gulf Coast, situated between Clearwater and Cape Sable, as a subregion to further identify this delineated study area.

Florida has 412 miles of Gulf shoreline or barrier beaches. Another 739 miles of coast are classified as bay or estuarine beaches. The beaches of the barrier islands serve as the primary line of defense against heavy weather events that would otherwise damage the upland or estuaries. Hurricane and tropical storm wave energies are lessened by the existence of barrier islands and their dynamic beaches.

The estuarine systems of the subregion consist primarily of tidal flats and mangrove forests. Like the outer barrier beaches, the intertidal and subtidal flats serve a buffer function. These bottoms provide a reproductive and developmental habitat for a variety of shellfish, fishes, some reptiles, and benthic marine invertebrates. Other wildlife forms, such as shore and wading birds, utilize these tidal zones for feeding and loafing.

The vastness of the mangrove forests of the subregion is legend. These too serve the upland as buffers against severe storms and as critical nursery areas essential to the life history of important marine resources. Commercial and recreational fishing, including shrimping, are interrelated to the mangrove community in terms of productivity. Over sixty species of juvenile fish are associated with, or dependent on, the mangrove estuaries for varying periods of their life cycle. There are approximately 390,000 acres of invaluable mangrove forest on the Florida Gulf Coast.

The barrier island beaches of the subregion are of a somewhat discontinuous configuration. They developed early in the Holocene period, or about 10,000 years ago. Most are classified dynamically as moderate energy beaches, with an offshore ramp slope of two to three feet per mile. Inshore water depths within a half mile of the barrier islands are generally about twenty feet deep.

The outer islands of Tampa Bay and Charlotte Harbor are the most extensive in terms of barrier beach development, as opposed to the beaches near Venice and Naples which are almost contiguous mainland.

The barrier beaches, especially Sanibel Island, consist of soil materials that have been transported from offshore, or were produced during a time of lower sea level when erosion of Tampa Bay and Charlotte Harbor resulted in soil distribution seaward. Approximately ninety percent of the sediments that resulted in the development of Sanibel Island about 5,000 years ago, are of biogenic origin. This demonstrates the geological youth of Sanibel and its east-west orientation as a result of storms and southward littoral drift.

The elevation of the great majority of the beaches within the subregion is low profiled, subject to inundation by hurricane surge and characterized by poor dune development. The absence of well elevated foredune systems probably accounts for primarily monotypic nesting which is restricted to the loggerhead turtle. Two other marine turtle species, the green turtle and leatherback, rarely nest on the Florida Gulf Coast and both of these species usually create a body pit prior to egg cavity excavation. This deeper nest site development requires a higher beach profile to avoid egg clutches being situated too close to the water table. Well-elevated locations ensure drainage during flooding which can result because of higher than normal tides or heavy precipitation.

Population migration in the United States has resulted in portions of Southwest Florida having the dubious honor of being identified, at least once in recent years, as one of the fastest growing sections in the nation. Along with this surge of human population numbers over the past one and a half decades, there has also been accelerated and haphazard coastal development lacking proper planning or adequate guidelines to ensure even a semblance of order. Once remote islands have been linked to the mainland with bridges or causeways, and a few others are awaiting the same fate. Human pressures on the barrier islands and the surrounding marine-estuarine systems continue to degrade the very essence of these habitats. Public recreational use, coastal construction, pollution, and the quantities involved in these environmental impacts, exert negative influences on the land and marine organisms whose success depends on the viable integrity of the very ecosystems that are being stressed.

The Gulf beaches along the coast of Southwest Florida are very closely associated vegetatively. A few of the plant species which occupy the Gulf

Plate 1. A pristine subregional nesting beach on Cayo Costa Island, Lee County, Florida. *Casuarina* had not invaded this natural beach system which is dominated by sea oats, sea grapes, and cabbage palms.

beach ridge and foredune ecosystems play important environmental roles, while others may negatively impact sea turtle nesting.

Soils or sand particles that are moved onshore by water currents and tidal action continue to flow across the berm of the dry frontal beach, carried along by the wind. Immediately upland from the spring tide line, the pioneer plants, mostly grasses and vines, begin to flourish and entrap the mobile sand particles being transported by the wind and, thereby, help to stabilize the formation of dunes. Primary plant species occupying the pioneer zone of the barrier beaches within the subregion include: sea oats *(Uniola paniculata)*, saltmeadow cordgrass *(Spartina patens)*, dune panic grass *(Panicum amarulum)*, railroad vine *(Ipomoea pes-caprae)*, beach elder *(Iva imbricata)*, camphorweed *(Hetrotheca subaxillaris)*, sea purslane *(Sesuvium maritimum)*, and inkberry *(Scaevola plumieri)*. As dune systems develop, age, and stabilize with organic materials provided by the decomposition of the open beach vegetation, woody shrubs and trees soon dominate the mature Gulf beach ridge system. Native, subtropical plants, many of them among the group known as West Indian hardwoods, are the climax species of the back beach. Among the subregion's intermediate and upper plant species are: Spanish bayonet *(Yucca aloifolia)*, wax myrtle *(Myrica cerifera)*, sea grape *(Coccoloba uvifera)*, prickly-pear cactus *(Opuntia stricta)*, wild coffee *(Psychotria undata)*, gray nickerbean *(Caesalpinia bonduc)*, bay cedar *(Suriana maritima)*, cabbage palm *(Sabal palmetto)*, sand live oak *(Quercus virginiana geminata)*, southern red cedar *(Juniperus silicicola)*, gumbo limbo *(Bursera simaruba)*, and the introduced exotic Australian pine *(Casuarina equisetifolia)*.

The *Casuarina* occupies virtually every lineal foot of Gulf beach ridge on the barrier islands from Tampa Bay to Cape Romano. Only beachfront that has man-made structures has escaped the proliferation of this invading tree. This one species of foreign tree continues to ruin sea turtle nesting beaches on both coasts of southern Florida. On Sanibel and Captiva as well as other nearby barrier islands, it marches in tiered canopy elevations, commensurate with age, ever closer to the water's edge. Almost every native beach plant in its path is out-competed by this noxious interloper which was introduced on the lower Florida peninsula in the early 1900's.

2
Sea Turtle Biology: An Overview

Turtles belong to the vertebrate class Reptilia, the subclass Anapsida, and the order Testudines. This order has been further subdivided into two suborders—Pleurodira and Cryptodira. The Pleurodira are a divergent group known as the side-necked turtles. These creatures literally retract their heads by bending them sideways, or laterally, as opposed to straight in, or vertically, like turtles of the Cryptodira. Side-necked turtles occur on or adjacent to three continents—Africa, Australia, and South America. North American turtles are members of the Cryptodira. The world's sea turtles also belong to the Cryptodira and are further split into two families.

When one thinks of turtles, usually the first feature that comes to mind is their unique shell. The armor-like outer covering essentially has remained unchanged throughout time. The skeletal bones of the turtle shell are numerous and complex. The bone arrangements comprise the carapace, or upper shell, and the plastron, or lower shell. The top and bottom sections are joined together by a structure called the bridge. An overlay of horny scutes typically covers the bony shell. These do not align with the skeletal bones beneath them and are fewer in number. There are, however, exceptions to this hard outer shell characteristic. Shells of the members of the families Trionychidae (soft-shelled turtles), Carettochelyidae (pig-nosed turtle), and Dermochelyidae (leatherback turtle) differ since their external shell covering lacks horny scutes or laminae. Figure 1 identifies the skeletal bones of a typical loggerhead carapace and Figure 2 indicates the arrangement of the scutes of the upper shell in the same species.

Modern sea turtles have highly modified extremities that have developed because of the animals' aquatic lifestyle. The phalanges of the limbs have become extended to form thin, flattened, oar-like flippers which gain optimum resistance to the water during a swimming stroke. The anterior flippers can move them through the water column gracefully and with astounding speed, and the rudder-like manipulations of the smaller posterior flippers lend maneuverability. When a turtle is ashore for reproductive

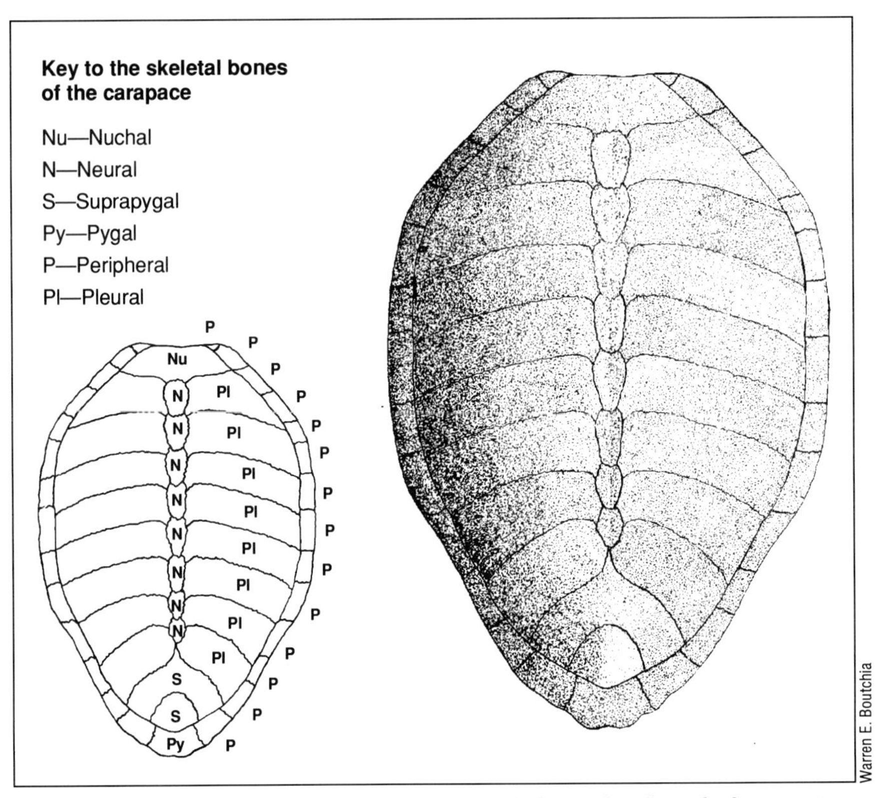

Figure 1. The skeletal carapace of an adult male loggerhead turtle from waters near Sanibel Island, Lee County, Florida. Carapace length (straightline)—37.5 inches (95 cm), carapace width (straightline)—27.75 inches (68 cm).

demands, the rear limbs are used with amazing dexterity in the excavation of the egg chamber.

All modern turtles are toothless and their jaw surfaces are covered with a one-piece keratinous covering known as a tomium; however, the fossil record indicates that a few of their antecedents possessed teeth. The jaw surfaces of marine turtles, depending on species, serve as crushing, cutting, or retaining devices. In the genera *Caretta* (loggerhead), *Lepidochelys* (ridley), and *Eretmochelys* (hawksbill) the powerful mouthparts are utilized for crushing shellfish, or reef-dwelling organisms, upon which these turtles prey. The more specialized jaws of *Chelonia mydas*, the green turtle, are used for cutting or shearing the submergent marine vegetation upon which they graze. The upper edges of the tomium of the green turtle lower jaw are serrated and coincide with the smaller serrations on the cutting edge of the upper jaw. The interior of the latter tomium is

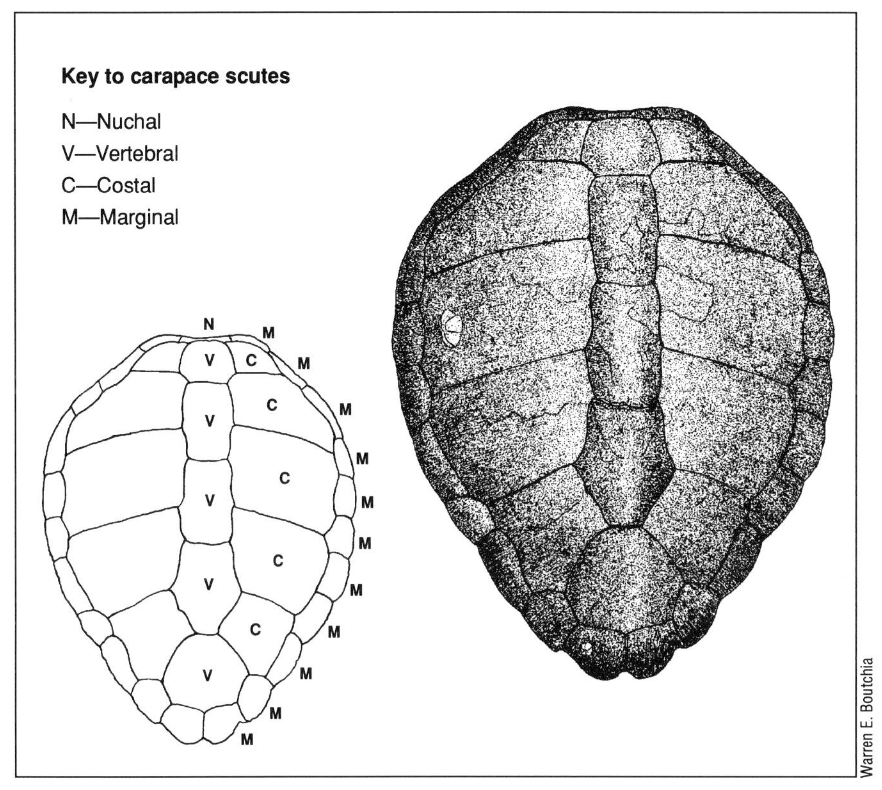

Figure 2. Carapacial scute arrangement of an adult female loggerhead from waters near Sanibel Island, Lee County, Florida. Carapace length (straightline)—36 inches (91 cm), carapace width (straightline)—30 inches (76 cm).

grooved to further accommodate the shearing action. In contrast to these similar jaw styles, that of *Dermochelys* (leatherback) is very different. The soft bodied animals that are the components of the leatherback diet are easily caught and held in position during consumption because of an adaptively developed skull feature. Its jaw coverings are not horny, but rather flexible in relation to the restricted soft diet of the species. One additional feature that makes the leatherback's mouthparts so unique is the existence of a pair of large cusps located at the front of the upper jaw. This strange anatomical feature is as close to teeth as any recent turtle species has had. The cusps, along with a deep recess in the center double premaxillary bones and a high indent laterally in the anterior portion of each of the maxillary skull bones, provide well-developed retaining devices. Figure 9, on page 20, illustrates the head of *Dermochelys*.

Sexual dimorphism in marine turtles is evident in sexually mature individuals. The prominent external sex characteristics of male sea turtles are: the development of specialized hooked foreflipper claws (with the exception of *Dermochelys* for this animal lacks claws) and the length of the tail, which extends a considerable distance beyond the posterior end of the carapace. In comparison, the claws of the female do not enlarge and the tail does not extend beyond the carapace. There are other, more subtle, sexual differences in sea turtles; for example, in *Caretta*, the shell shape is different between the sexes, and the head features and general coloration of the male, in my experience, appear to be dissimilar from the more commonly observed female of the species. The carapace of male loggerheads is more elongated and narrower, proportionately, than that of females, and their head structure seems different in a profile orientation than that of the opposite sex.

In turtles, the female oviducts, rectum, and urethra enter the cloaca. This is a cylindrical chamber which empties to the exterior of the animal via the anus. A clitoris, which is homologous to the penis, is also present as a thickening of tissue on the ventral, or bottom, wall of the cloaca. In the male turtle, the cloacal chamber also receives body wastes and, in addition, contains the penis. The male sex organ is situated in the ventral muscular wall of the cloaca. Penile eversion is accomplished by vascular engorgement of spongy tissue which projects from each side of the bladder duct on the anterior cloacal wall. Prior to mating, erection of the penis results as the tissues fill with blood. Penetration is accomplished and fertilization is internal. Spermatozoa flow from the testes to the female cloaca and enter the two separate oviductal orifices via a penis that has a grooved seminal channel which is unlike the mammalian penis that houses an extension of the enclosed urethra as a seminal conduit.

At ovulation, sea turtle ova penetrate the membranous wall of the ovary, enter the coelom, and travel on to the ostium of the oviduct. They are then moved along through the system by ciliary and muscular action. During their passage through the oviductal tube, other components of a maturing egg are added. Albumen is secreted in transit to envelop each egg and, later, specialized glands add the thin shell wall as the eggs reach the lower section of the oviduct. In order for fertilization to occur, sperm must reach the egg before shell surfaces are developed. The inner surface of the oviduct possesses narrow bands of cilia which move towards the ostium and assist sperm in reaching eggs at the crucial time.

Like all chelonians distributed over the earth, sea turtles are inextricably tied to the land for their reproductive needs. Females must visit dry beaches where they deposit quantities of round pliable-shelled eggs. The copulatory act usually occurs a few weeks earlier offshore. All species apparently reproduce on a kind of individually controlled cycle that is not completely understood. Some species nest every two years, while others nest on three or even four year intervals; however, individuals of some species have been recorded nesting in successive years. There are individual females who have switched back and forth from two to three year intervals, and then back to a two year cycle again.

Each female will visit land at regularly spaced intervals for nesting. These multiple egg depositions are maintained on an internesting interval timetable which can range between ten and twenty nights. Individuals of some species, notably *Chelonia*, have been recorded as producing a maximum of up to eleven clutches during one nesting season, but very few female green turtles lay eggs so profusely.

The eggs of turtles are altogether different than those of birds and will be discussed later in detail. Some turtles produce round eggs with a hard and brittle calcified egg shell, i.e., the gopher tortoise *(Gopherus polyphemus)*, while others may lay elongated or soft eggs. All marine turtles deposit round, pliable eggs with parchment-like shell surfaces. The shell casing is permeable and allows unrestricted passage, on demand, of water and gases required by the embryo for proper development. There are major differences in the volume ratio of shell, white, and yolk between a hen's egg and a sea turtle egg. The chicken egg, by weight, is approximately 11 percent shell, 56 percent white, and 33 percent yolk. In contrast, the typical sea turtle egg consists of about 5 percent shell, 40 percent white, and 55 percent yolk.

The interior of a sea turtle egg is made up of several layered elements or extraembryonic membranes. The outermost component is the chorion and it is this membrane that encapsulates the whole inner egg. The fluid-filled amnion is situated within the chorion and will provide moisture essential to successful embryological development. The allantois, located near the interior of the amnion, is the major respiratory organ and waste reservoir. This connects to the hindgut of the turtle embryo. Another membrane forms around the yolk to create the yolk sac. These complex membranous elements are surrounded by the aqueous albumen which is retained by the shell wall. In fertile sea turtle eggs a spherical membrane, the vitelline sac, or embryonic disc, surrounds the embryo, rises to the top of the egg, and adheres to the inner surface of the shell wall.

The senses of sight, smell, taste and hearing are developed to varying degrees in sea turtles. Their eyes have adapted for utilization beneath the water surface and because of this, sea turtles are myopic when ashore and out of their medium. They are very sensitive to light and movement; in fact, there is considerable evidence that turtles have color perception, and they may utilize this sense in identification of preferential foodstuff.

The overall level of development of their sense of smell is a complex issue of sea turtle behavior and biology; it is not fully understood and is the subject of limited disagreement between marine turtle biologists. Many

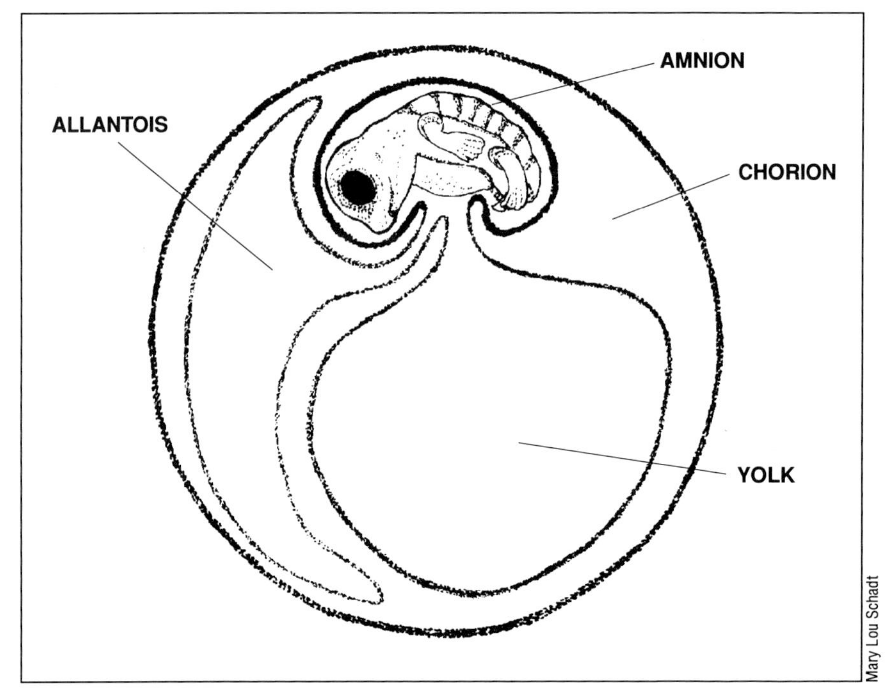

Figure 3. A sea turtle egg in cross-section.

specialists in turtle behavior are convinced that important olfactory functions occur in these animals. The behavior by female turtles of a few species, as they probe and push the beach sand with their snout shortly after they make landfall on the upland edge of the wet beach for nesting, is believed by some marine turtle specialists to be olfactory in nature. Other biologists are of the opinion that this behavior is simply a physical reaction of a turtle being out of its aquatic element. In other words, the observed behavior may be simply the result of an individual resting its heavy head as it pauses during the strenuous ascent of the beach surface. Heavy breathing, which usually accompanies these pauses, may then not be related to olfactory considerations, but coupled to periodic rest cycles because of the exhausting trip up the beach.

The ability of turtles to taste articles of food is considered to be poorly developed. Some turtle biologists are convinced food selection is based on sight and smell, with taste having no genuine consequence in the selection of an individual's dietary items. However, this does not explain why loggerhead turtles raised in captivity develop definite taste preferences that are seemingly cyclic. Some individuals will eagerly accept specific food for two

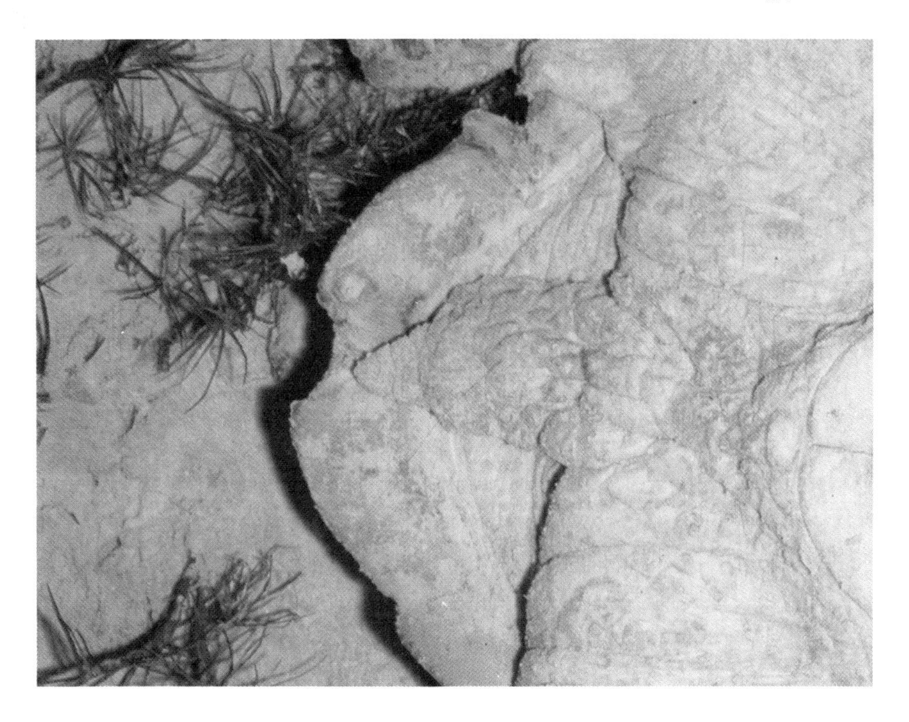

Plate 2. The tail tip of an adult female loggerhead does not extend beyond the rear of the carapace.

or three weeks; then, they will suddenly reject it and prefer some other material in their diet for a few weeks.

Turtles do not have external ears, but they do have internal auditory receptors. Their hearing system consists of a middle ear and an internal ear. It is known that they can hear, but at what decibel level waterborne or airborne sounds are recognized, or perceived, is not fully understood.

Vocalization in sea turtles is extremely limited. The most boisterous of the group is the leatherback turtle. There have been anecdotal reports of *Dermochelys* which, while under attack by predators, emitted sounds which ranged from wails to roars. All sea turtles make a variety of sounds during nesting. These noises include deep sighs, grinding together of the jaws, and loud hisses when they are disturbed. Hatchling loggerhead turtles make strange, low volume, squeaking sounds while still inside the nest or when they are placed in containers and transported to the water's edge to be released.

Since sea turtles are confined exclusively to the marine environment, their bodily functions have developed the mechanics to purge their system

Plate 3. The tail of adult male loggerheads extends a considerable distance beyond the rear margin of the carapace.

of excess salts. This effectively allows them to internally convert saltwater to fresh water. Turtle kidneys, unlike those of marine mammals, are incapable of voiding excess salt which is dissolved in their blood. A lacrymal, or tear gland, is located in each eye orbit and constantly secretes a copious salty fluid. This occurs while the turtle is at sea or during the female's brief visit to a nesting beach. Without this specialized gland to supplement kidney functions, sea turtles would be unable to remain hydrated and balance their sodium and potassium levels.

Land and fresh water turtles are commonly known as poikilotherms, or cold-blooded animals. More correctly, they are ectotherms, for they do have primitive means of thermoregulation. Their body temperature is influenced by environmental elements, such as ambient air or water temperature, and chelonians regulate personal temperature by certain behavioral actions. Basking provides an increase in their core temperature, elevating it above that otherwise established by environmental considerations, and helps to maintain bodily functions at an optimum level.

Marine turtles have vaulted ahead of this basic thermoregulatory regimen and can advantageously control their body temperature. Because of its bulk, fat stores, and specialized vascular system, the leatherback turtle has the most advanced thermoregulatory structure of all the reptiles. This turtle can maintain its body temperature as much as twelve degrees above that of the seawater in which it swims. Body mass in the other marine turtles assists them in similar maintenance of their body temperatures. Sea turtles also bask on the surface of the water to gain as much sunlight radiation as possible and to assist in elevation of their core temperature.

Turtles can hold their breath and remain underwater for extended periods of time. In well-oxygenated water, some freshwater species can remain submerged indefinitely. Members of the family Trionychidae, the soft-shelled turtles, can take oxygen directly from the water through special tissue in the cloaca and throat. The ability to remain underwater in most aquatic turtles is related to the degree of individual activity and ambient water temperature.Cold water temperatures reduce the animal's metabolic demands down to a minimal level and their submerged time can be extended to hours or even weeks if the hibernation threshold is crossed. The same turtle, in warm water and in a high activity mode, could remain submerged for only minutes. There is also mounting evidence which indicates that stressed sea turtles, those threatened by hypothermia or asphyxiation due to forced submergence, can rely on anaerobic metabolism for a short period of time.

All sea turtles can descend to considerable depths, but the record-holder for deep diving is the leatherback turtle. Investigators in the U.S. Virgin Islands have attached time-depth recording instruments to post-nesting leatherbacks. These devices recorded the dive depths and durations for two turtles. One of the individuals, during the ten day test period, dove to an average depth of 222 feet. This animal made a maximum dive to 1,558 feet which lasted for 27.8 minutes. Another leatherback dove to a depth of more than 3,280 feet. The actual depth was beyond the capacity of the instrumentation and the dive took 41 minutes. Readings from the depth recorders were compiled when the female leatherbacks returned for renesting at the end of their internesting interval.

The leatherback turtle is the record holder for breath-hold deep dives. Sperm whales have been recorded sounding to nearly 3,000 feet, and man can breath-hold free dive to about 100 feet.

3
Sea Turtle Lineage

There are varying levels of disagreement among paleontologists and herpetologists as to what early prehistoric reptilian life form constituted the original stock from which the modern turtles descended. The earliest animals which can be classified as turtles belonged to the family Proganochelyidae and occurred in present day Europe during the Triassic period, or about 230 million years ago. The primitive reptiles of this family were probably marsh dwellers. Later, offshoots from these stem turtles selected various habitats and, today, two families, the Cheloniidae and the Dermochelyidae represent the modern and exclusive group of true marine turtles.

The Cheloniidae evolved from the fundamental marshland Proganochelyidae, and its lineage very early opted to enter and utilize the marine environment. By the lower Cretaceous period, this family, represented by primitive *Caretta*-like ancestors, had evolved to near its present level of development. By the upper Cretaceous period, about eighty million years ago, the genus *Caretta* had reached its present level of development. Representatives of today's other marine turtle genera of the Cheloniidae—*Chelonia*, *Lepidochelys*, and *Eretmochelys*—are known from the Eocene period or later.

Earlier in this century, some sea turtle specialists were of the opinion that the other extant sea turtle family, Dermochelyidae, had developed from some early marine reptile apart from the ancestral stock of the Cheloniidae. In other words, they suggested that while the Cheloniidae may have developed in a stepping stone fashion, sea-land-sea, the Dermochelyidae fulfilled evolutionary responses entirely in the sea without earlier near-terrestrial counterparts. Today, this is not considered to be an accurate hypothesis. Modern authorities are now convinced that the Dermochelyidae also originated from a hard-shelled ancestor like Cheloniidae. In modern times, this family is monospecific and contains only one genus, *Dermochelys*.

Despite being quite different morphologically, the two families have many similarities biologically. These are mostly behavioral likenesses and are discussed later in the appended section on allied species.

Two sea turtle families are extinct. These are the Toxochelidae and the Protostegidae. The latter family contained the largest turtle known to have ever lived—*Archelon ischyros* which ranged over mid-America when the land of today was ocean bottom. The fossil remains of this prehistoric species have been collected from upper Cretaceous deposits in the United States. The carapace length of this giant could reach twelve feet with the animal attaining a weight of two tons.

During this lengthy evolutionary process, extraordinary physical changes occurred. To streamline the sea turtle configuration, the proportional shell size was reduced and its shape altered; consequently, today's marine turtles have lost the benefit of retracting their head and appendages into their shell for self-protection. This was an advantageous trade-off to attain speed and maneuverability.

Although modern sea turtles are represented by two living families, the Dermochelyidae and the Cheloniidae, there have been attempts at further subdividing the members of the Cheloniidae into subfamilies or tribes. Some authorities place the genera *Chelonia* and *Natator* in the tribe Chelonini and the genera *Caretta, Lepidochelys* and *Eretmochelys* into the tribe Carettini.

A further breakdown of these existing families and their tribes, genera and species is given below.

Family CHELONIIDAE

Subfamily (Tribe) Chelonini

Genus and species	*Chelonia mydas*	Green turtle
	Chelonia agassizi	Black turtle
	Natator depressa	Flatback turtle

Subfamily (Tribe) Carettini

Genus and species	*Caretta caretta*	Loggerhead turtle
	Lepidochelys olivacea	Olive ridley
	Lepidochelys kempi	Kemp's ridley
	Eretmochelys imbricata	Hawksbill turtle

Family DERMOCHELYIDAE

Genus and species	*Dermochelys coriacea*	Leatherback turtle

4

The Sea Turtles of the Gulf of Mexico

Of the eight species of marine turtles, five are known to occur in the Gulf of Mexico in modern times: the green turtle, the loggerhead turtle, the Kemp's ridley, the hawksbill turtle, and the leatherback turtle. Another species, the olive ridley *(Lepidochelys olivacea)*, ranges through the Indo-Pacific system and is also found along the northeastern coast of South America, in the Guianas. This ridley was extending its non-nesting range north from that region until as recently as a decade or so ago. Today, this species is widely exploited by man, and indications are that it is in decline in the Western Atlantic; however, it is conceivable, though a remote possibility, that one day this wide-ranging species may reach the Gulf of Mexico.

I consider each of the above forms to be world-wide monotypic species, unless restricted in distribution by historical range as in the case of Kemp's ridley, and none as having subspecific forms anywhere. Eventually, there may be additional taxonomic splitting, or further subdivision, of stocks of *Chelonia mydas* and, ultimately, valid subspecies of the green turtle may be described.

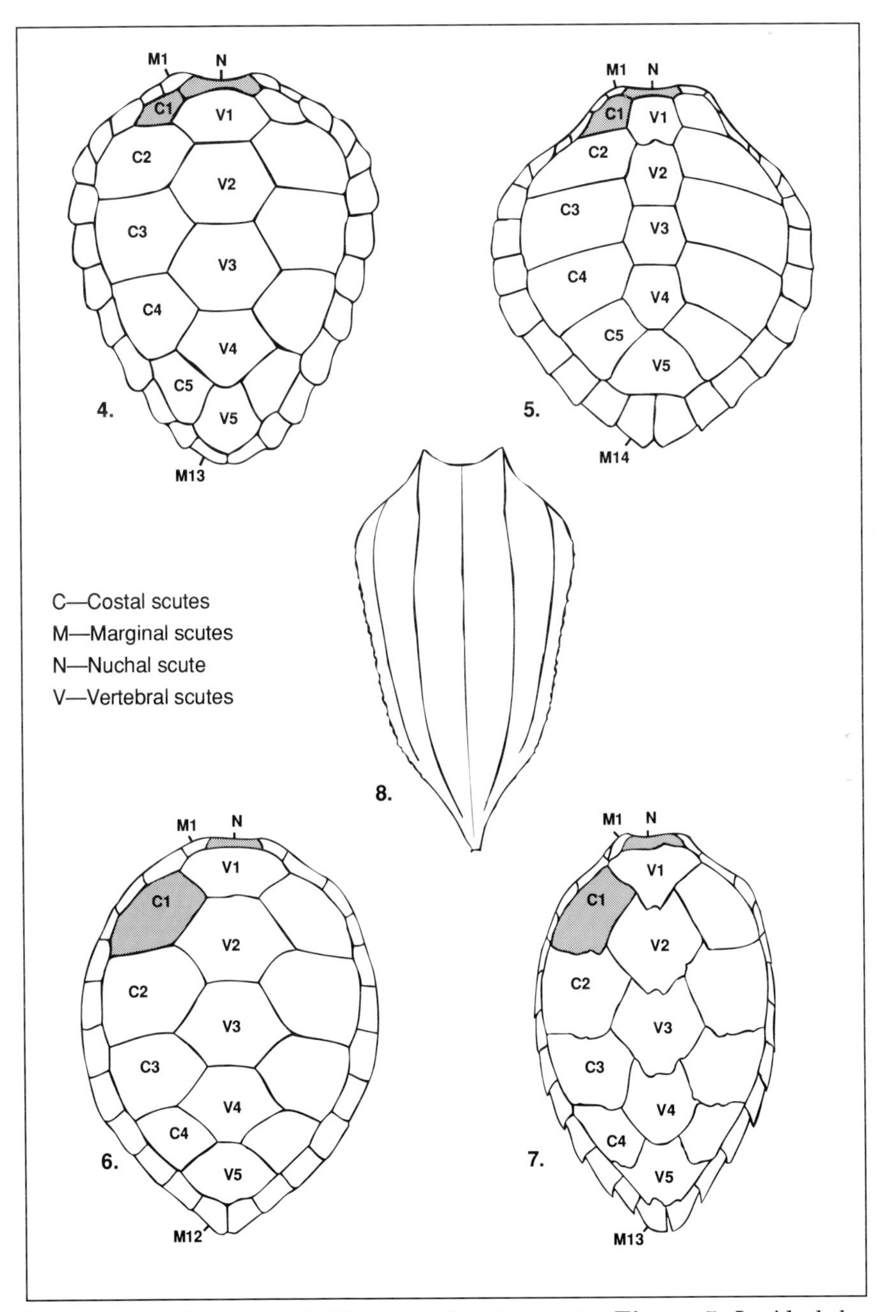

Figures 4-8. Carapaces of: **Figure 4**, *Caretta caretta;* **Figure 5**, *Lepidochelys kempi;* **Figure 6**, *Chelonia mydas;* **Figure 7**, *Eretmochelys imbricata;* **Figure 8**, *Dermochelys coriacea.*

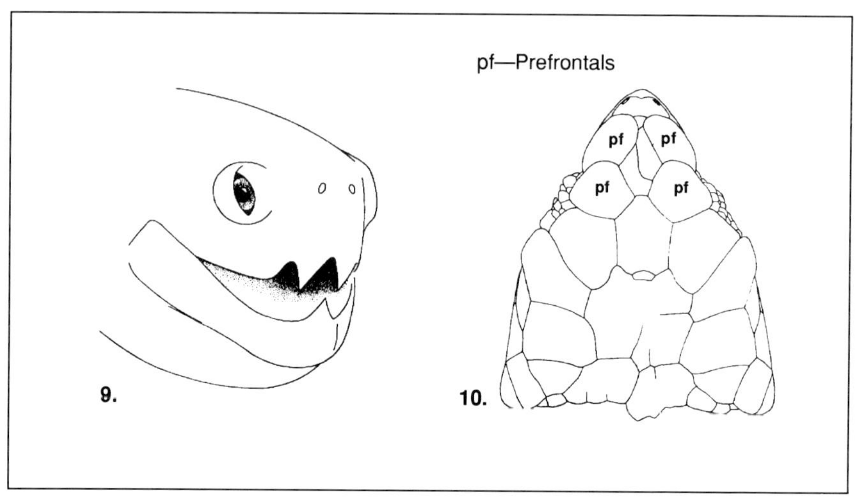

Figures 9-10. Figure 9, *Dermochelys coriacea;* **Figure 10**, *Caretta caretta,* scutes on top of head.

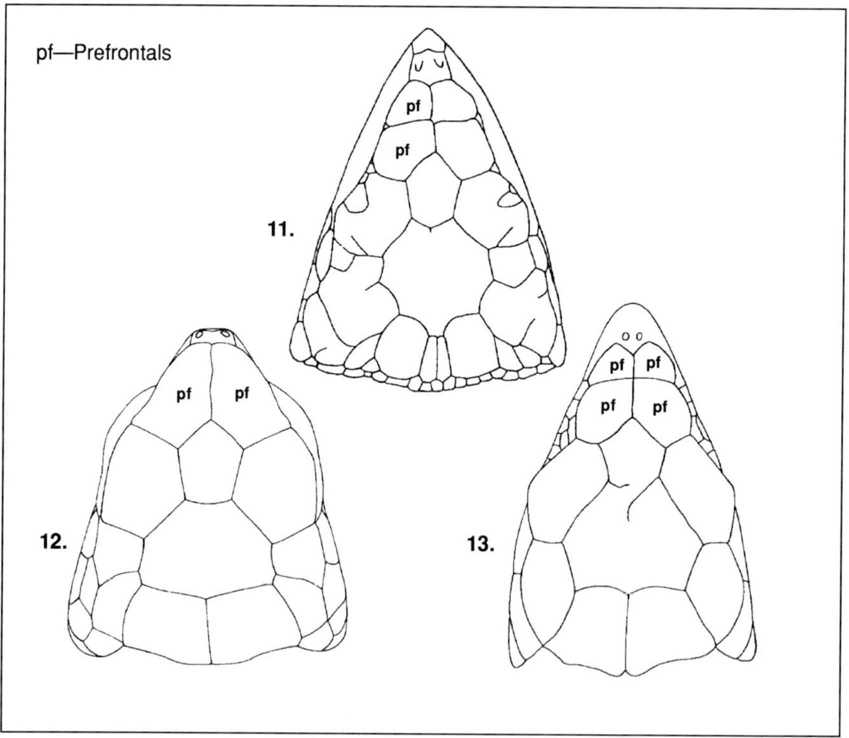

Figures 11-13. Scutes on the upper head surface: **Figure 11**, *Lepidochelys kempi;* **Figure 12**, *Chelonia mydas;* **Figure 13**, *Eretmochelys imbricata.*

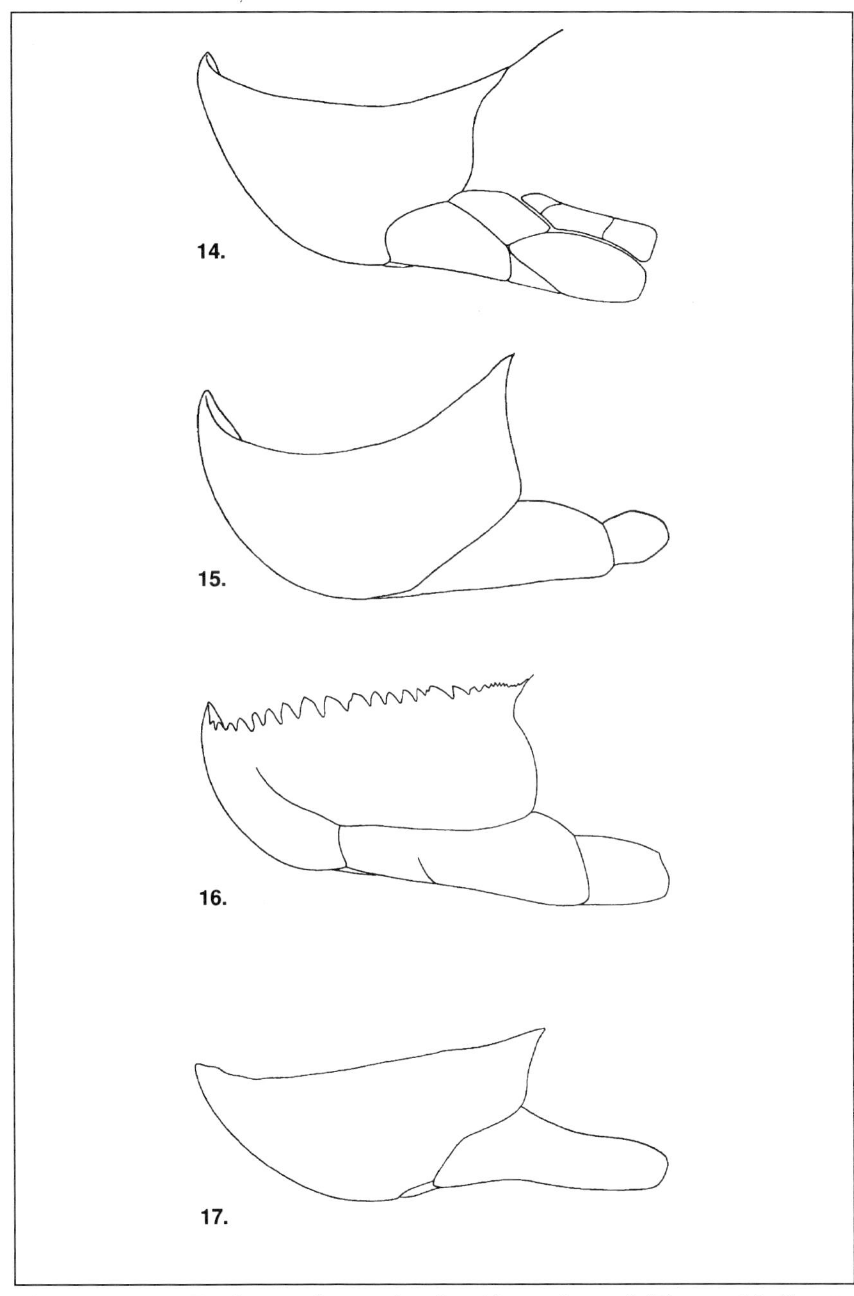

Figures 14-17. Tomium, or horny sheath, of lower jaws of: **Figure 14**, *Caretta caretta;* **Figure 15**, *Lepidochelys kempi;* **Figure 16**, *Chelonia mydas;* **Figure 17**, *Eretmochelys imbricata.*

Part 2

5 The Loggerhead Turtle
6 Food Habits
7 Tags and Tagging
8 Reproduction
9 Non-nesting Emergences
10 Nest Site Tenacity
11 Fecundity
12 Egg Size and Shell Thickness
13 Clutch Sizes
14 Incubation Period
15 Embryonic Development
16 Egg and Hatchling Mortality
17 Beyond the Nest
18 Beach Selection
19 Adult Mortality

Caretta caretta

5
The Loggerhead Turtle

Carolus Linnaeus (1707-1778), a Swedish naturalist, is responsible for laying the foundation of modern systematic zoology. His great work, *Systema Naturae,* for the first time applied the binomial system of nomenclature, consisting of genus and species, to plants and animals. In the tenth edition of his classic work (1758) he named a turtle from America which had five scutes on its back—*Testudo caretta.*

Today's taxonomists identify this animal as the marine turtle *Caretta caretta* (Linnaeus), the loggerhead turtle. The development of this species' scientific or taxonomic name has been a progression of change from the work of Linnaeus to the present. Over the past two hundred years the loggerhead has been known by the following synonymous scientific identities.

Testudo caretta Linnaeus 1758
Testudo caouana Daudin 1802
Caretta nasuta Rafinesque 1814
Thalassochelys caouana Fitzinger 1833
Caouana cephalo Cocteau 1838
Thalassochelys caretta Boulenger 1889
Caretta caretta Siebenrock 1909

Taxonomists at one time divided the genus *Caretta* into two races or subspecies: *Caretta caretta caretta,* the Atlantic loggerhead and *Caretta caretta gigas,* the Pacific loggerhead. The invalid differential characters which once divided the two forms primarily were based on the number of marginal scutes edging the carapace and the skeletal bones of the upper shell. Subsequently, researchers have shown that the number of these scutes and bones are variable among the different geographical stocks of the species and do not justify division of *Caretta* into subspecific forms.

The glossarial definition of the loggerhead turtle's scientific name is (caretta), a tortoise shell.

The common, or vernacular, name of the loggerhead turtle is used consistently in the English-speaking part of the globe. *Caretta* is properly

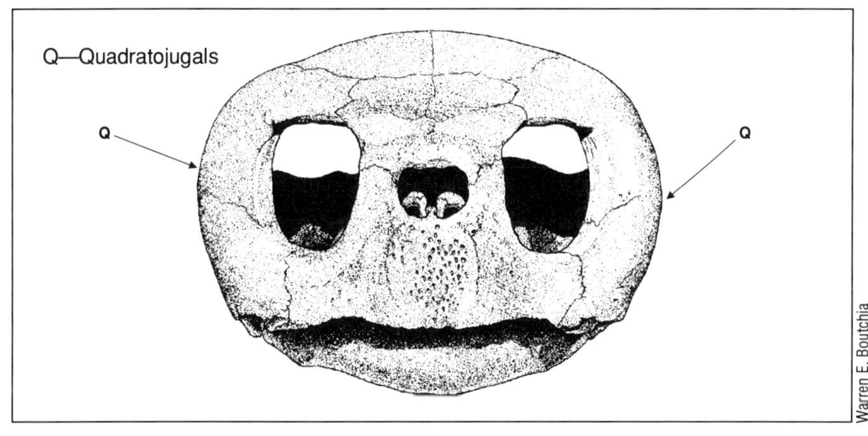

Figure 18. Frontal view of the skull of an adult male *Caretta caretta* from waters near Sanibel Island, Lee County, Florida.

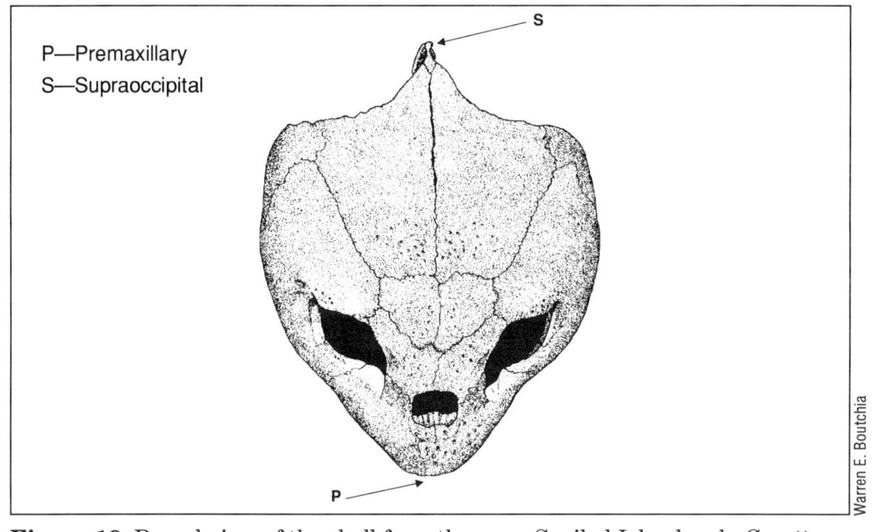

Figure 19. Dorsal view of the skull from the same Sanibel Island male *Caretta.*

called loggerhead because it possesses an enormous, somewhat out of proportion, cranium. Illustrations of the skull structure of an adult loggerhead are shown on pages 26, 27, and 28. The length of the individual's skull, from which measurements were taken with the tomium removed, from the premaxillary to the supraoccipital, is 32.5 cm. The skull width, also straight-line measurement, between the quadratojugals, is 25.8 cm. The cranial bones between which the above measurements are made are labeled in Figures 18 and 19.

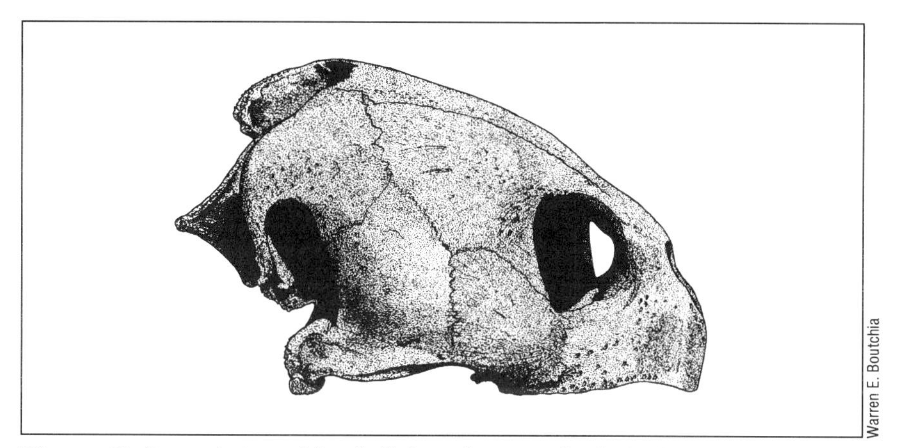

Figure 20. Lateral view of the skull from the same Sanibel Island male *Caretta*.

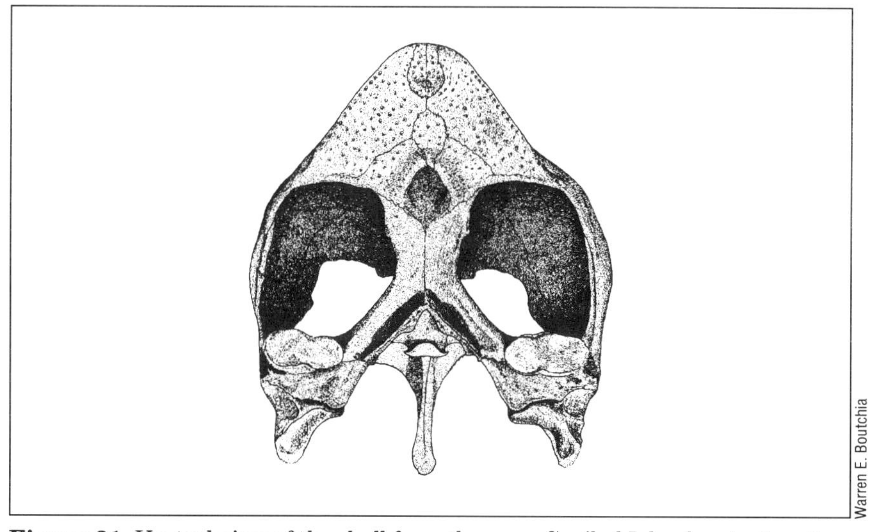

Figure 21. Ventral view of the skull from the same Sanibel Island male *Caretta*.

The loggerhead's upper shell pattern is composed of five costal and five vertebral scutes. Delineation of the scutes in adult *Caretta* is difficult since most mature loggerheads have the carapace covered by a multitude of benthic fouling organisms—from dense algaes to enormous barnacles that may approach three inches in diameter. Some mature individuals that I have examined over the years have had sections of their carapacial scutes lifting away from lower keratinous layers. This condition may be a consequence of sloughing of the outer scute strata as a result of age, injury, or poor adhesion. Separation may also be caused by growth pressures of

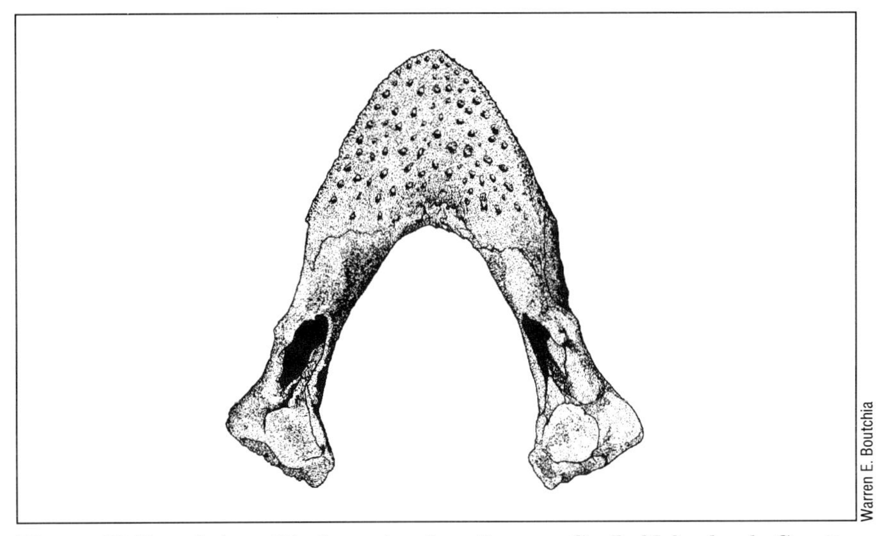

Figure 22. Dorsal view of the lower jaw from the same Sanibel Island male *Caretta*.

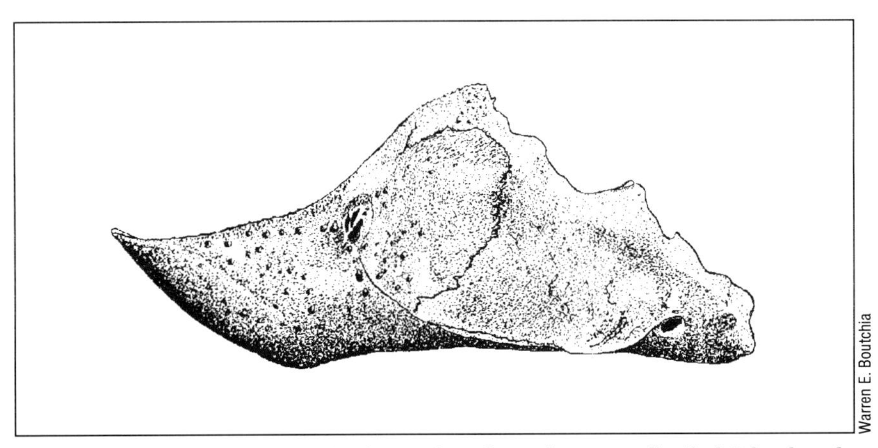

Figure 23. Lateral view of the lower jaw from the same Sanibel Island male *Caretta*.

otherwise harmless benthic commensals that displace the thin layers from underlying tissue.

The adult loggerhead's skin is reddish-brown and light yellow in color with the epidermis of occasional specimens having yellow hues which blend into a more orange than red base color. The carapace, when not obscured by benthic fouling organisms, is a reddish-brown, but may have radiating hues of olive melded into the base color. The lower shell and bridge structure connecting it to the carapace are usually cream-colored. The base color of the male skin appears to be more of a brown tone than the red-orange of the female.

Hatchlings vary widely in coloration within the same clutch. The carapace ranges from a light beige to mahogany brown to almost jet black and is darker along the median than in the marginal areas. The plastron also ranges in color among siblings and may be anywhere from a light beige to a dark gray. Hatchlings with dark pigmented carapaces usually are correspondingly dark on the venter. More drastic colored hatchlings, ranging from albinism to melanism, are produced occasionally.

Commensurate with aging, considerable alteration occurs to the shell configuration of the loggerhead turtle, with respect to the length/width ratio and the shape of its scutes. At hatching, the carapace has three dorsal keels and the plastron has two ridges, all of which are situated longitudinally to the shell. These keels and ridges disappear as the individual grows, but low knob-like projections remain on the rear of each keel through the juvenile size class. On attainment of sexual maturity, at a straightline carapace length of near thirty inches, the keels and knobs have disappeared and the carapace has become generally smooth.

The marginals posterior of the bridge in subadults and the smaller examples of recently matured young adults appear well defined. The suture between each of these scutes is indented, giving a serrated appearance when the shell is viewed from above. As the individual grows, these indentations become obscure, and their borders are no longer well defined, except for the cleft created at the point of union of the two posterior-most marginals.

Almost all of the adult loggerheads that I have handled in the course of my studies typically have had their carapaces nearly covered by a variety of benthic organisms. Colonization by these marine plants and animals occurs at different density levels and is probably indicative of the water depth or quality in which an individual turtle resides. Most loggerhead carapaces, and to a lesser extent, their heads, are heavily encrusted; however, I have observed a limited number of specimens which possessed rather beautiful upper shells, for they were virtually free of these epibionts.

Other species of marine turtles, notably *Chelonia* and *Dermochelys*, are usually free of adhering organisms as adults. The lifestyle of *Caretta* and the characteristics of its shell probably contribute to the usual luxuriant invertebrate and algal colonization. Loggerheads customarily sleep on the bottom and situate themselves tightly against rocks in cave-like rocky depressions, reef outcrops or wrecks, or they dig into and conceal themselves in the bottom sediment. Some of the animals which I have found in the carapacial community of these turtles certainly moved into this microecosystem under their own locomotion. Others must have arrived as spat, or at some other time during their larval stage.

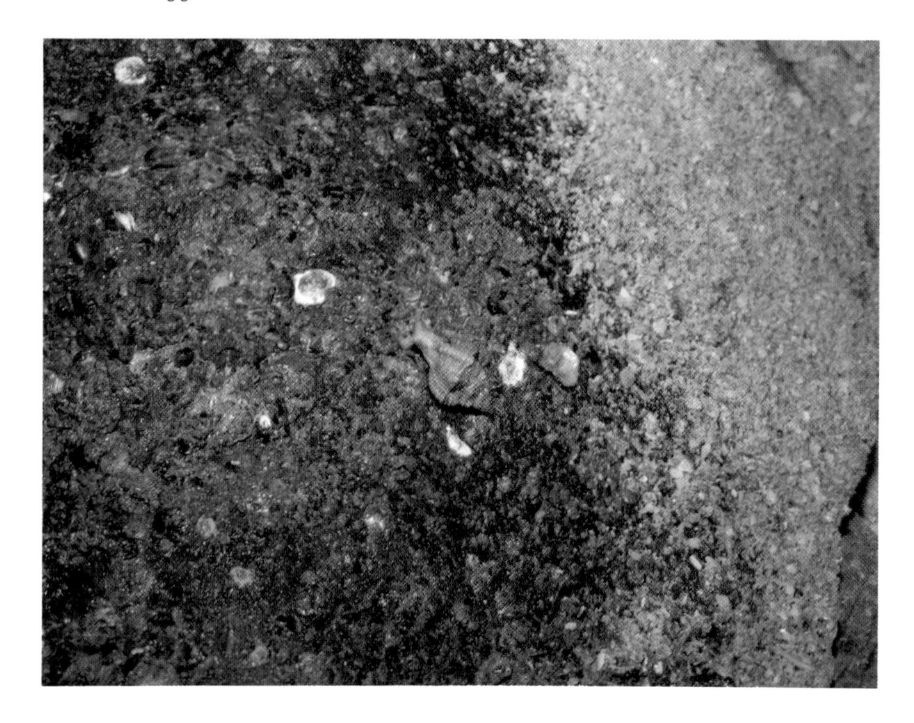

Plate 4. A typical *Caretta* carapace epibiont community. Scores of species of organisms regularly foul their shells. One function of the loggerhead turtle in the marine ecosystem is its role as a transporter species: it carries invertebrates across deep water and major currents, obstacles that may otherwise restrict their distribution. An apple murex, *Phyllenotus pomum*, is in the center of the photograph.

Studies of the fouling organisms that are a part of the loggerhead turtle carapacial benthic community have been made along the Atlantic seaboard of the United States. It has been suggested that there may be two nesting populations of loggerhead turtles along the Atlantic coast. This is based on the existence of two distinct carapace communities which were identified, based on epibiont composition, as a northern and a southern population. These two populations would equate to two different nesting populations and do not include transient individuals who have moved out of their nesting habitat and may be temporarily travelling through either the northern or southern epizootic habitats.

For several years, I collected carapace organisms from nesting loggerheads within the subregion. Collections were made coincidental to a related study aimed primarily at identifying barnacle species which colonized the shell of *Caretta*. The barnacle material was preserved in five percent seawater formalin and, in 1976, I forwarded a collection to Henry Spivey who was then associated with the Department of Biological Sciences at Florida State University, Tallahassee. He identified eight species of barnacles and found one additional form I had collected from subregional *Caretta* that had not been previously scientifically described.

To my knowledge this barnacle, shown below as *Platylepas* n.sp., was never described in the scientific literature. At the time, Henry Spivey requested that I collect a larger series of the animal for his taxonomic analysis, but I was never successful in collecting an adequate quantity.

The barnacle specimens examined contained the following species:

Lepas anatifera anatifera
Lepas anserifera anserifera
Chelonibia caretta
Chelonibia patula
Balanus trigonus
Balanus venustus
Platylepas hexastylos
Platylepas n.sp.
Stomatolepas elegans

All of the above are known from the Gulf of Mexico. Both species of *Lepas* are cosmopolitan in distribution. *Chelonibia caretta* is found in all tropical and subtropical seas, while *C. patula* occurs in the tropical and subtropical Atlantic, Mediterranean, and Indo-West Pacific. *Balanus trigonus* is a cosmopolitan species found principally in tropical and subtropical waters. *Balanus venustus* is distributed in the Atlantic and Indian Oceans, and has been reported from Japan. *Platylepas hexastylos* is known in all tropical and subtropical seas from the skin of turtles, manatees, and dugongs. *Stomatolepas elegans* is a cosmopolitan barnacle attaching to the soft skin and throat of the marine turtles *Caretta*, *Chelonia*, and *Dermochelys*.

Adult loggerhead turtles range up to nearly four feet in carapace length. The early literature reports weights of up to 1,200 pounds. Loggerheads of such great weight are probably a thing of the past—if indeed they ever existed. The weights of adult female loggerhead turtles estimated by casual observers have generally been grossly exaggerated. The early published accounts of maximum weights of *Caretta* were sometimes based on skull dimensions which have no real relevancy to body size or weight. By closely monitoring physical data produced by my long-term tagging program, it is evident that calculation of a turtle's body weight based on carapace size is far from accurate. When the recorded carapace dimensions are reviewed for those turtles which return to subregional beaches with biennial or other cyclic regularity, there seldom is noticeable growth in shell

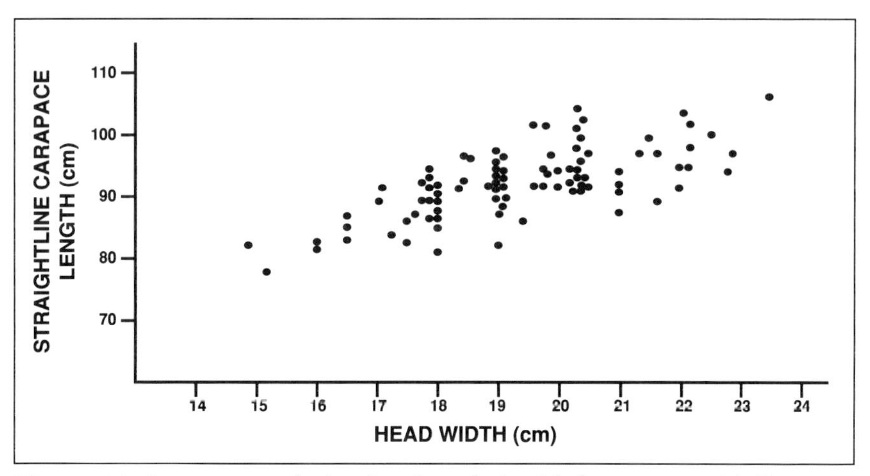

Figure 24. A scatter-graph showing the relationship between straightline carapace length and head width in nesting loggerhead turtles from Sanibel Island.

measurements that remove the bias of possible human error. Skull dimensions, however, are another matter. Specimens which show little or no carapacial growth usually have experienced an increase in the dimensions of their cranial width.

I have not attempted to correlate shell length or width, or head size to body weight. In my opinion, there have been far too many variables in the limited sample size available to make an accurate judgment. Variations in head size between individuals of near equal carapace length, absence of one or more appendages, individual obesity, and even the extent to which benthic organisms colonize the carapace make any accurate correlative determination difficult, if not impossible.

Female loggerhead turtles were weighed on Sanibel Island during the nesting seasons of 1972 through 1976. Fifty-nine post-nesting turtles were weighed in this study. A Chatillon 500 pound capacity hanging scale was utilized. A large folding tripod was erected on the beach, and from this support a small block and tackle assembly was suspended. The scale was connected to the block and tackle and the final section to be added was a modified piece of shrimp trawl netting. Turtles to be weighed were carefully turned upside down onto the net, the tripod apparatus was positioned above them, and the turtle was hoisted until clear of the beach.

Recorded weights ranged from 175 to 355 pounds. The lightest individual, CR 392, measured 34 inches (86.4 cm), straight carapace length (scl) and 24.5 inches (62.2 cm), straight carapace width (scw) and had a straightline (sl) head width of 7.5 inches (19.1 cm). At the upper end of the range, CR 1007, had a scl of 39.5 inches (100.3 cm), a scw of 30 inches (76.2 cm), and

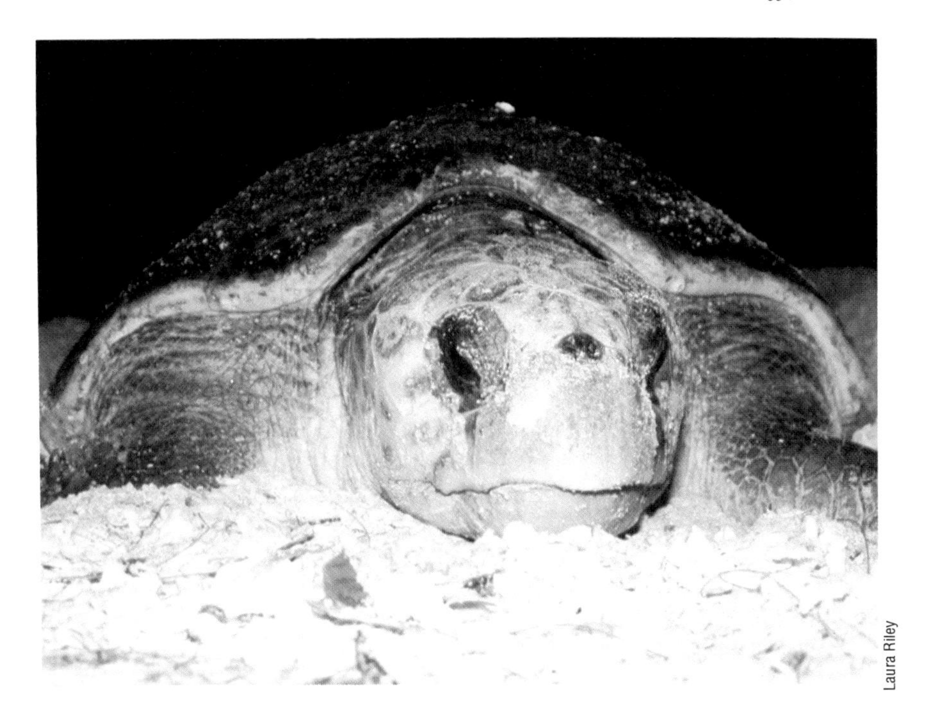

Laura Riley

Plate 5. The loggerhead is well-named. Its massive head attains dimensions greater than that of any other living turtle.

a sl head width of 8.75 inches (22.2 cm). Turtle CR 392 was missing approximately one half of her front left flipper. The mean weight of the fifty-nine loggerheads sampled was 247.7 pounds.

Over the years, I have turned over many large loggerhead turtles. When one learns the basic technique, it is not difficult and does not require great strength. In 1988, we encountered a large loggerhead which had just made a false crawl and was about to enter the surf. Eve Haverfield and I leaped out of the Jeep and, in an attempt to restrict the turtle's movement and escape, tried to invert her. We did not have time to grab our turtle restraining device. Although Eve has turned loggerheads which weighed close to 300 pounds, both of us strained to get this ponderous animal upside down, and we were unable to do so. Exhausted, we watched helplessly as the giant turtle moved ahead and disappeared into the dark sea.

The crawl this turtle had left on the wet beach was over 113 centimeters in width. Eve appropriately named this turtle "Moby Dixie." Through the remainder of the nesting season we continued to search the beach on subsequent patrols, hoping to cross her path again. We measured each wide

Plate 6. A loggerhead situated in the weighing apparatus and about to be hoisted clear of the beach.

crawl after that experience, but never found one to equal that of "Moby Dixie." This was, indeed, the largest loggerhead turtle I had ever observed and, although I am not fond of estimating weights, I would judge this animal's weight to be in excess of 500 pounds.

When the head width of specimens with near identical carapace dimensions is examined, it is further evident that there is no correlation between shell length and head width, relative to body weight. Turtle CR 1007 was larger in scl than turtles CR 162 and CR 389. These individuals measured 36.25 (92.1 cm) and 36.5 inches (92.7 cm) respectively. CR 162 had a head width of nine inches (22.9 cm), while CR 389's head measured 8 inches (20.3 cm) across. CR 162 weighed 265 pounds and CR 389 weighed 245 pounds.

Male loggerhead turtles from the Gulf of Mexico adjacent to Sanibel Island are slightly larger than females from the same waters. Although opportunities to measure living males have been almost nonexistent, there are measurements available from stranded specimens. A small sample of sixty-two individuals, thirty-one of each gender, is available from strandings in 1987, 1988, and 1989. The mean straightline carapace lengths for

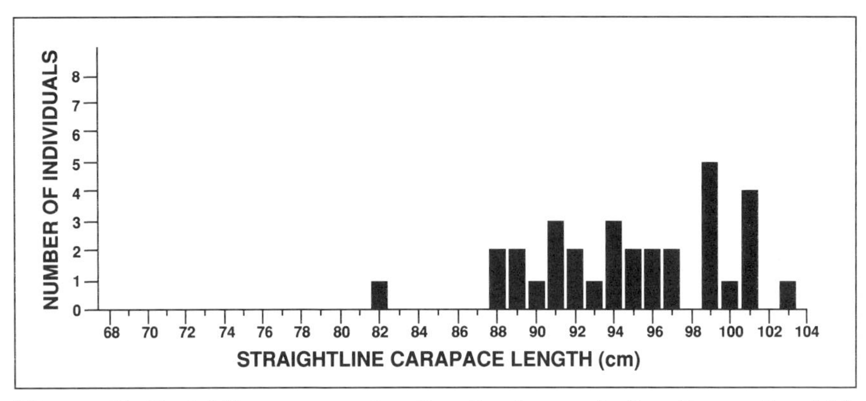

Figure 25. Straightline carapace lengths of mature male *Caretta caretta* which stranded on Sanibel Island. 1987-89.

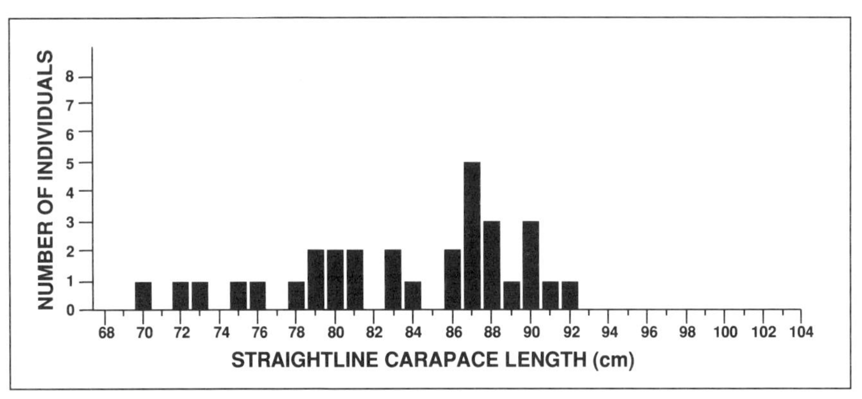

Figure 26. Straightline carapace lengths of mature female *Caretta* which stranded on Sanibel Island. 1987-89.

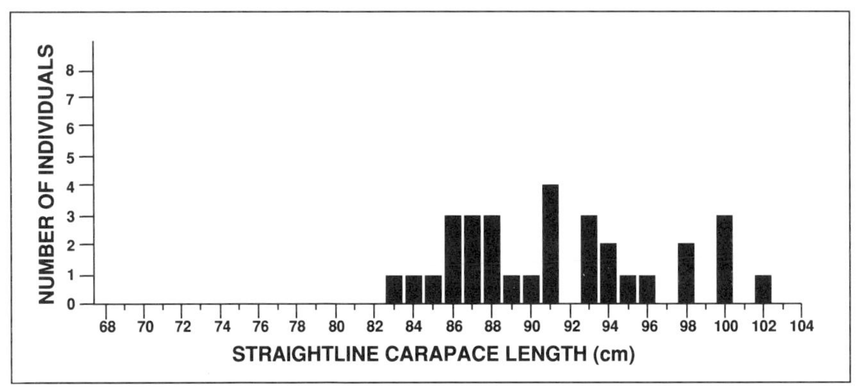

Figure 27. Straightline carapace lengths of untagged female *Caretta* nesting on Sanibel Island. 1987-89.

these two groups of dead animals are: males—37.4 inches (95.0 cm) and females 32.7 inches (83.3 cm). Only individuals of sexually mature size ranges or those possessing dimorphous external sex characteristics were placed in the sample.

When the above figures are compared to measurements collected from nesting females on the Sanibel Island beach, there is a rather remarkable difference in the average size dimensions between stranded and live females. Nesting females, from a sample of thirty-one new or untagged turtles, averaged 35.6 inches (90.4 cm) in straightline carapace length. The differences in size between the two groups—stranded animals that are not part of the nesting population and reproducing females—will be discussed later.

At hatching, there is some variability in the size of little loggerheads in both carapace dimensions and total weight. In 1973, neonate turtles, which were produced in a predation-offset transplantation effort, were measured (carapace length only) and weighed. Measurements were straightline and taken at the same physical location at which carapacial lengths are determined in adults—between the nuchal, anteriorly, and the outer-most rear marginal, posteriorly. Individuals measured were free of any external yolk remnants and their carapaces appeared to be completely extended. If the shells of the hatchlings did not appear to be straight, they were separated and did not become part of the sample measured. Weights were recorded in grams with an Ohaus triple-beam balance utilized for weighing. Hatchlings were dry and only weighed following careful brushing to remove all sand particles.

One thousand six hundred eighteen hatchlings were measured. This represented turtles from a total of twenty clutches. Individual sizes ranged from 38.0 to 50.0 mm, with a mean carapace length of 46.65 mm for the sample.

The number of hatchlings weighed was more restrictive. Because of the time element which was required to dry and clean each individual, the weight sampling only included ten specimens randomly selected from each clutch weighed. Weights ranged from 16 to 22 grams, with an average weight for the group of 19.6 grams.

Hatchling turtles produced by a loggerhead turtle, CR 140, who will be discussed in detail later, provided an interesting overview of sibling relationships, in terms of clutch size, egg, and hatchling sizes. These data are shown in Table 1.

The majority of modern herpetologists who specialize in the marine turtle group recognize *Caretta caretta* as a monotypic form occupying much of the earth's oceanic habitat. It is found in the Atlantic, Pacific, and Indian Oceans, as well as contiguous, but lesser water bodies.

Each of these major ocean shores have significant and very large nesting assemblages of *Caretta caretta;* for example, exceptionally dense

Clutch Number	Date Deposited	Clutch Size	TOP Egg Size (in mm)	BOTTOM Egg Size (in mm)	Hatchlings Produced	Range of Carapace Length (in mm)	Average Carapace Length (in mm)
73-5	5/24/73	174	41.2 × 41.9	41.5 × 42.3	137	46.5–48.0	47.19
73-9	6/4/73	162	38.0 × 38.8	39.4 × 39.4	128	40.0–48.0	45.78
73-15	6/18/73	152	39.5 × 40.3	39.6 × 40.4	130	43.0–49.0	47.18
73-29	7/5/73	145	38.5 × 39.4	39.2 × 40.0	121	43.0–47.5	45.38
73-32	7/15/73	144	41.4 × 42.3	41.8 × 42.4	117	45.5–50.0	47.45
73-41	7/29/73	143	40.6 × 41.8	40.9 × 41.8	112	44.0–49.0	47.06
	Total	**920**			**745**		

Table 1. Clutch size, mean egg dimensions from a sample of ten top and ten bottom eggs per clutch, hatching success, and carapace lengths of hatchlings produced by loggerhead turtle CR 140. Sanibel Island, 1973.

reproducing populations of the loggerhead exist in a six-county area on the east coast of Florida, on Wreck Island in Queensland, Australia, and on Masirah Island, Oman. Less profuse populations utilize the Indian Ocean beaches of Tongaland, South Africa, the Pacific in southern Japan, the Atlantic coasts of Colombia, barrier islands of South Carolina, and in Turkey on the coast of the Mediterranean.

It is estimated that there are somewhere around 30,000 mature female loggerhead turtles ranging through available habitat and utilizing the various nesting beaches along the western Atlantic. The term "western Atlantic" used herein refers to the southeastern United States, and to a much lesser extent, some beaches close to the Caribbean. Without question, the great percentage of these mature turtles use beaches in Florida, Georgia and South Carolina.

The nesting range limits of the loggerhead turtle in the United States are situated between New Jersey on the Atlantic seaboard, south toward Florida, and west to Texas. Nesting by the species is infrequent north of Virginia and north of Tampa Bay in the eastern Gulf of Mexico. Historical records of the loggerhead turtle nesting to the west of the Mississippi delta are scant. In late June, 1969, a loggerhead emerged from the Atlantic onto Staten Island, New York, but was frightened away by curious onlookers who happened to be on the same beach. The northernmost valid nesting record for the species in the United States is Atlantic City, New Jersey.

Beyond the limits of its nesting range, the loggerhead's distribution continues northward to the Canadian Maritime Provinces. Periodically, live or stranded loggerhead turtles are found in Great Britain or on the adjacent continental European coast. These turtles probably originate in American

Figure 28. Historical North American nesting range of *Caretta caretta*.

waters and are transported overseas by the North Atlantic Current via staging from the Gulf Stream.

The loggerhead turtles of the subregion are wide-ranging in their non-nesting distribution and move great distances between nesting seasons. Tag returns, or tag number notifications, have provided considerable new information on postnesting dispersal, or the general whereabouts, of females after they have been tagged on their respective nesting beach. I would prefer to have tags remain attached to the turtles and be notified of the tag number, date of an animal's capture or observation, and location. Unfortunately, such ideal circumstances, for continuity of data, have seldom been the case. Most tag returns which I have received over the years have resulted because of loggerhead mortality.

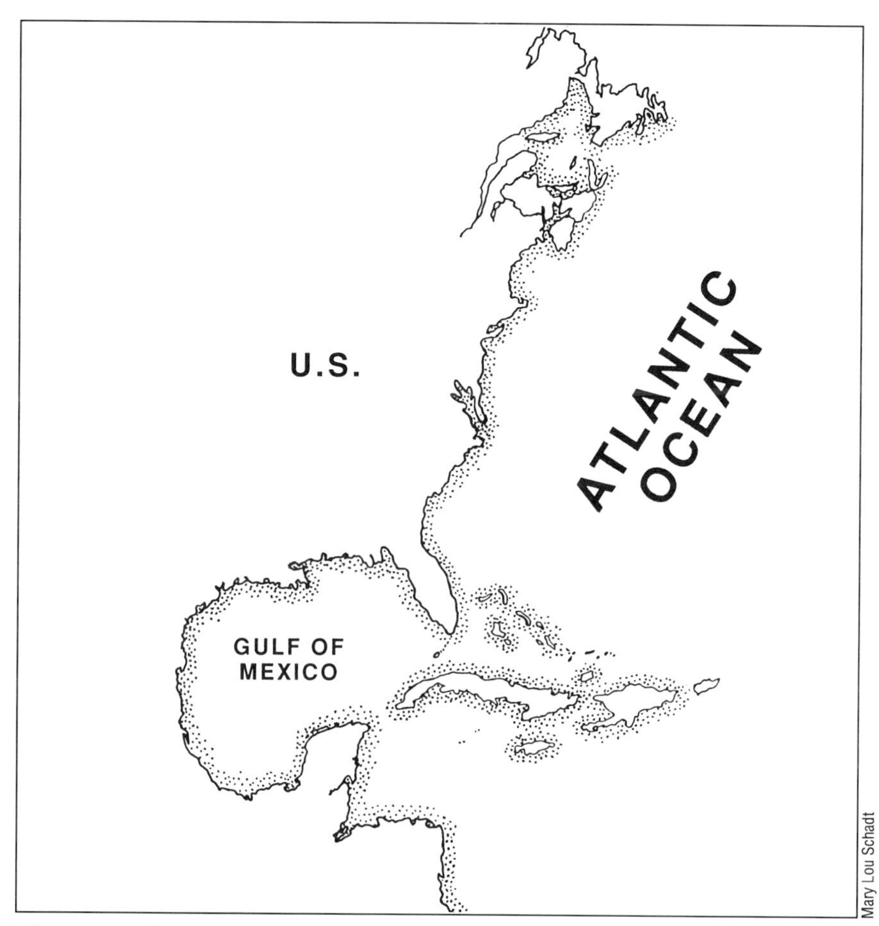

Figure 29. General distribution of *Caretta caretta* in the western Atlantic Ocean.

The following locations of tag recoveries are in no way a complete itemization of tag returns. They are meant to illustrate the non-nesting distribution of Sanibel Island loggerhead turtles based on recovery of tags. Within the scope of the tagging program of Caretta Research, Inc., the Sanibel Island tagging program is the most continuous, and the majority of tag recoveries from subregional loggerheads have originated from turtles tagged there. Because of this continuity, only data from *Caretta* tagged on Sanibel are utilized as references; however, dozens of other tags from turtles tagged elsewhere in the subregion have been returned from points within the area encompassed by these specific geographical locations.

Tag recoveries from the Gulf of Mexico have been the most frequent. A loggerhead tagged on Sanibel Island in 1969 was caught and drowned in

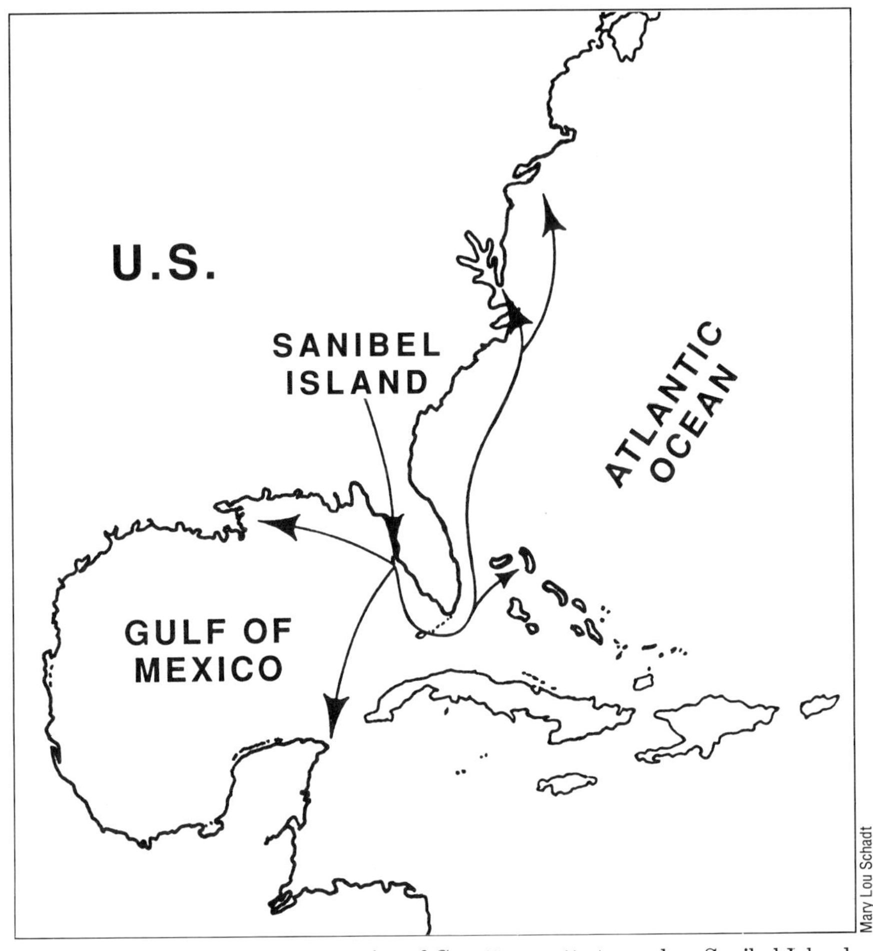

Figure 30. Long distance recoveries of *Caretta caretta* tagged on Sanibel Island. 1964-1989. Arrows are not intended to suggest routes traveled.

a shrimp trawl fourteen months later off the Chandeleur Islands of Louisiana. Another individual was netted by a Mexican fisherman at Cabo Catoche, just south of the tip of Mexico's Yucatan peninsula.

Information about one of my tagged turtles reached me from the vicinity of Cayo Fragoso on Cuba's northern coast. The letter which informed me of the turtle's capture came through a Miami importer's office. However, I never received the tag or additional information of the turtle's capture or demise. At least one other Sanibel loggerhead crossed the Gulf Stream. This individual was killed by a native Bahamian fisherman from Moore's Island, near Abaco, Bahamas.

The long-distance record holder for a subregional specimen is a Sanibel Island loggerhead which was caught and drowned in a pound net, approximately thirteen miles offshore, northeast of Atlantic City, New Jersey. If this individual paralleled the coast during her passage, she travelled in the vicinity of 1,500 miles from the point of tagging in just over eleven months.

One of the most unusual captures occurred in 1976 when an elderly sportfisherman caught one of my Sanibel Island turtles on a baited hook and line. This event took place in the freshwater York River, in Virginia, which empties into the western side of Chesapeake Bay. The tag was removed, the turtle released alive, and the tag mailed to me.

A very interesting tag recovery from a loggerhead turtle, quite unrelated to my work in Southwest Florida, occurred "down under." This turtle was tagged on an island of Australia's Great Barrier Reef. Sixty-two days later, this same turtle was recaptured 1,200 miles north of the point of tagging. This was not straight line open ocean travel, but rather, like the Sanibel turtle discussed above, the distance this specimen covered is a projected total of coastal miles from the point of tagging to the location of later recapture. To cover such a great distance, this remarkable loggerhead had to average nearly twenty miles of ocean travel per day.

The loggerhead turtle population of the subregion, in terms of total numbers, is difficult to ascertain. Since field work on the beaches has only included contact with mature females and hatchlings, other than dead stranded turtles, the composition and total number within the respective ranges of all size classes is unknown.

During the nesting seasons of 1970 through 1975, I conducted an aerial survey along 102 miles of the Southwest Florida coastline. The survey area extended from Morgan Pass, Collier County, Florida, north to Venice Inlet, Sarasota County. The purpose of this survey series was to determine: (1) which beaches were most frequented by nesting *Caretta;* (2) at what time period was nesting activity at the highest level; (3) did nesting peaks vary from year to year; (4) were loggerhead turtle nesting populations increasing or decreasing; and, (5) what changes in nesting patterns, at specific beaches, occurred from summer to summer. In 1976, Pat Hagan and I coauthored a paper on the role of this aerial survey in an attempt to estimate the subregional loggerhead turtle nesting population. This was presented at a Sea Turtle Conference that year. Some of the material used in the remaining pages of this chapter relates to this joint work.

Systematic aerial surveys of the Gulf beaches from Cape Romano north to Manasota Key were initiated on May 16, 1970, and ended on July 16 of that year. This survey study was terminated in 1975 and, in that year, flights were conducted from May 15 until July 20. The actual flights were executed at about two week intervals, usually on a weekend day to coincide with the personal schedules of volunteer pilots and aircraft availability. This frequency also helped to decrease the possibility of nesting crawls being counted twice.

Flights were made parallel to the various barrier beaches about 200 feet seaward and at an altitude of approximately 200 feet. Air speed varied from 80 to 100 miles per hour. Flights were scheduled at differing tidal stages and times of day. Pilots did not participate in the counting of nests; this was the responsibility of the one or sometimes two observer(s) aboard the aircraft. Observed crawls were considered a nesting crawl only when the track terminated in the customary dry beach zone above the line created by the most recent high tide, and clearly defined postnesting arched foreflipper sweep patterns were visible.

The observability of tracks and their identification as nesting or non-nesting emergences were hampered by several factors. Quick decisions on the part of the observer were required to place a loggerhead turtle crawl in the correct category. The decision reaction was difficult when crawls were close together, partially concealed by shadows, or partly washed away on narrow beaches. The effects of wind, rain and tide further limited observational accuracy. Strong winds from the west to northwest, in conjunction with periodic extreme high tides, often caused beach erosion. Other wind and current patterns often deposited accreted soils over crawl imprints. Severe weather events also moved marine vegetation on the face of the beach and concealed crawls. Human disturbances to some sections of the various beaches played a negative role as well and added to the concealment of nesting crawls, since heavy foot traffic and motor vehicle operation had obliterated tracks in some areas. Observational problems because of water glare and *Casuarina* shadows were minimized by flying the survey after 1200 hours.

In 1970, total nests counted from the air along the survey route tallied 759 and, in 1975, added up to 520. This reflected a 31 percent decrease in the number of total nests in the two seasons which were five years apart. It is not clear if these totals reflect a general population decline. Cyclic nesting periods for many subregional *Caretta* occasionally overlap. An overlap in nesting cycles can occur and result in an otherwise inexplicable surge in the number of nesting females visiting a beach. In other words, representative biannual nesting animals may be joined on the nesting beaches by a high percentage of loggerheads who are on a three year reproductive cycle. In 1970, the nesting population may have included an above average number of one and two year nesters who were joined by a significant number of turtles that were on a selected three year cycle. The total number of nesting loggerheads that year may have been inflated because of the overlap.

Coincidental to the crawl-counting overflights, ground-truthing nest counts were conducted on Bonita Beach, Sanibel Island, and Manasota Key.

These ground-truthing surveys were accomplished during the night or early morning prior to a scheduled aerial survey. Crawls which appeared to be the result of successful nesting landings were tabulated and compared with the results produced for each of the three beaches. Comparison between the two census techniques indicated that aerial nest counts tallied about 50 percent less than the total for the ground count. The ground-truth surveys allowed the observer to take the time to examine each crawl, determine its identity, and to better evaluate the many crawls which had terminated in the vegetated areas of the respective beaches.

To reach an estimated population level for adult female *Caretta,* the aerial counts were doubled to account for the disparity between air and ground counts. Using this methodology, the area included in the aerial survey study contained at least 1,518 loggerhead turtle nests in 1970 and 1,040 nests five years later in 1975. A subregional female loggerhead turtle makes several nesting landings each summer she reproduces, with three multiple nestings being average for each individual during her personally selected reproductive season. If this is considered and placed into the calculations, the subregional beaches surveyed had an estimated nesting population of 500 loggerheads in 1970 and 345 in 1975. These numbers do not provide any population estimates for subregional areas which are excluded because of the boundary established for the aerial work. I would venture to say that there are less than 100 female loggerhead turtles nesting on subregional beaches north of Venice Inlet, in any one summer. South of Morgan Pass, limited nesting occurs on a few of the larger and outer Ten Thousand Islands, dense nesting occurs on the beaches of Cape Sable, and some nesting occurs on the available beaches in Florida Bay, and the Florida Keys—Dry Tortugas. Probably fewer than 50 individual loggerheads nest on beaches of the Ten Thousand Islands and the other island areas combined. Approximately 250 turtles are reported to nest on Cape Sable beaches in good nesting seasons. When these totals are considered, I estimate that the adult female population of loggerheads, in the eastern Gulf of Mexico, contains in the neighborhood of 1,800 individuals.

An accurate sex ratio for mature loggerhead turtle populations is unknown. Studies in the Atlantic waters near Cape Canaveral have indicated a level of 1 female to 1.6 males. In Georgia, developers of life history models for *Caretta* which nest on Little Cumberland Island assume a 1:1 sex ratio for hatchlings produced on that beach.

Assuming that an equal 1:1 sex ratio exists in the adult subregional *Caretta* population, then a mature population can be further extrapolated from the above figures to arrive at a relatively reliable population estimate for the eastern Gulf of Mexico. This does not include or consider, however, the unknown numbers of a subadult population that exists in subregional waters.

For Sanibel Island, specifically, there has been considerable variation in the total number of nests produced each year. In 1959, when only nine miles of beach could be patrolled to the west of the Lighthouse because of the location of Blind Pass, that portion of the island's beach contained an estimated 220 loggerhead turtle nests. By 1971, when natural beach changes allowed extended vehicular coverage of 12.5 miles of the island's shoreline only 70 nests were tallied. In 1989, because of erosion on the western end of the island, 11.5 miles of beach were included in the monitoring program and the survey totalled 111 nests.

6
Food Habits

The diet of the loggerhead turtle is composed of a variety of marine organisms. Dietary components may consist of selected or preferred food items, while others may be secondarily ingested or consumed because of the opportunistic feeding habits of the species. In captivity the loggerhead will accept a wide variety of foods, including: various fishes, crabs, shrimp, squid, jellyfish and other marine animals; and prepared foods, i.e., Top Choice dog food, commercial trout feed; and, when offered, greens such as lettuce and spinach. Information available on the animal's natural diet in the wild has been obtained from frequent examination of the gastrointestinal systems of stranded turtles. The data which are provided by examination of the stomach and intestinal contents of stranded loggerhead turtles have been helpful in understanding their food preferences.

Loggerhead turtles prey on benthic organisms and other animals that move freely though the water column. The powerful jaws, with their well developed crushing surfaces, enable the turtles to nearly pulverize extremely hard and thick-shelled mollusks, and to pry or pull off other foods that are firmly attached to objects on the sea floor. Free-swimming jellyfish of several species are also consumed by the loggerhead turtle.

Jellyfish break down quickly because of the digestive processes; thus, little evidence of them is ever discovered during necropsy. There are many written accounts of the loggerhead turtle consuming the dangerous Portuguese man-of-war *(Physalia physalis)*. Most of these records for United States waters originate from the Atlantic coast since the man-of-war is not a common resident of the eastern Gulf of Mexico. The loggerhead turtle can consume this jellyfish without any discernible ill effects. The stinging cells of the man-of-war apparently do not cause severe internal toxic reactions because of the sea turtle's strong gastrointestinal constitution, or some physiological adaptation that reduces or eliminates immediate and long-term effects. An occasional blind loggerhead has been reported from areas having high populations of the man-of-war. In captivity, loggerheads which are fed these jellyfish close their eyes when consuming them. The eyes are

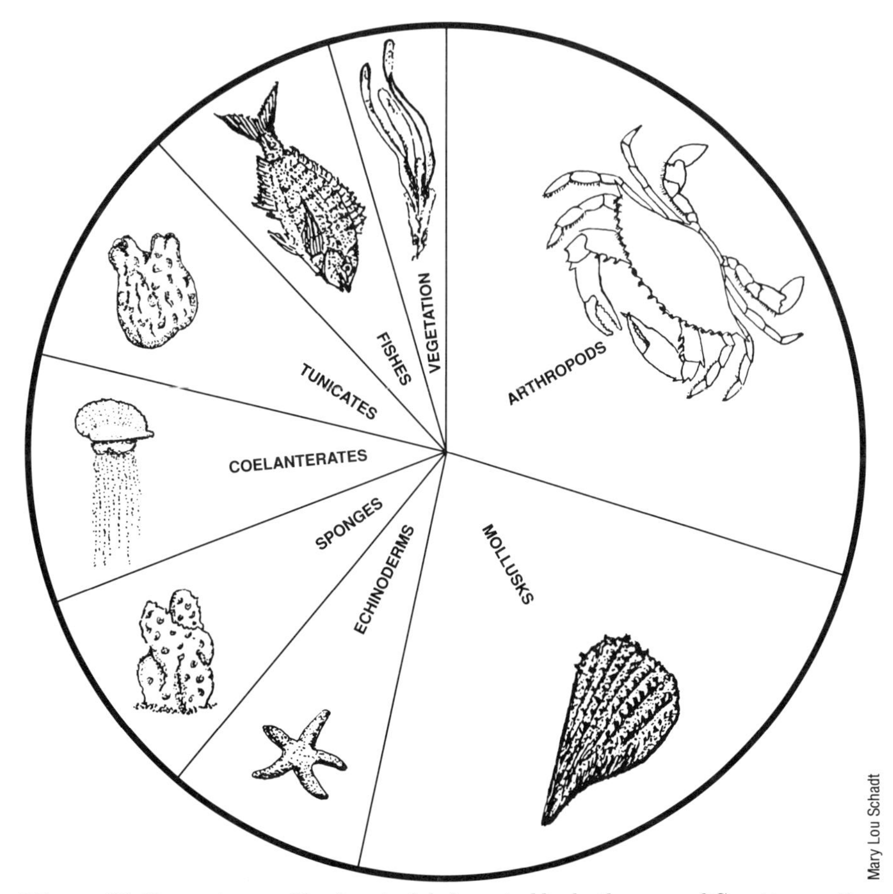

Figure 31. Percentages of food materials ingested by both sexes of *Caretta caretta* in the vicinity of Sanibel and Captiva Islands, Lee County, Florida. Partitions are based on the occurrence of identifiable foodstuff contained in the gut.

sensitive to the nematocysts, or stinging cells of jellyfish, and can be permanently damaged if not protected.

The food pie above illustrates the complex loggerhead turtle diet. The sections of this graphic are based on the analysis of gastrointestinal tract contents which were collected during the necropsies of over one hundred sub-adult and adult loggerhead turtles. The specimens which were examined stranded on the Gulf and bay beaches of Sanibel and Captiva Islands during the last three decades.

The following marine organisms were removed from the gastrointestinal system of an adult female loggerhead turtle which stranded on Sanibel

Island on April 10, 1974. The specimen was located nine miles west of the Sanibel Lighthouse on the Gulf beach. The prey species provide an example of one individual loggerhead turtle's food choices.

Stiff pen shell	*Atrina rigida*
Saw-tooth pen shell	*Atrina serrata*
Many lined lucine	*Lucina multilineata*
Comb bittersweet	*Glycymeris pectinata*
Cross-barred venus	*Chione cancellata*
Transverse ark	*Anadara transversa*
Coquina	*Donax variabilis*
Ponderous ark	*Neotia ponderosa*
Calico clam	*Macrocallista maculata*
Crested tellin	*Tellidora cristata*
Slipper shell*	*Crepidula fornicata*
Eastern white slipper*	*Crepidula plana*
Baby's ear	*Sinum perspectivum*
True tulip (operculums only)	*Fasciolaria tulipa*
Banded tulip	*Fasciolaria hunteria*
Apple murex	*Phyllenotus (Murex) pomum*
Lightening whelk	*Busycon contrarium*
Pear whelk	*Busycon spriatum*
Fighting conch	*Strombus alatus*
Paper fig	*Ficus communis*
Acorn barnacle*	*Balanus* ssp.
Worm shell*	
Oyster*	
Purple sea urchin	*Arbacia punctulata*
Stone crab	*Menippe mercenaria*
Giant hermit crab	*Petrochirus diogenes*
Small hermit crab	
Calico crab	*Hepatus epheliticus*
Flame-streaked box crab	*Calappa flammea*
Purse crab	*Persephona punctata*

*These species probably were attached to the exterior of *Strombus alatus* which were inhabited by hermit crabs.

It is surprising that this specimen did not contain the remains of horseshoe crabs *(Limulus polyphemus)* and blue crabs *(Callinectes sapidus)*. These crustaceans usually account for a considerable percentage of the loggerhead's food. Since this individual originated from the Gulf, perhaps the bottom conditions where this specimen had most recently foraged were not a preferred habitat for these crabs.

Plate 7. The stomach contents of a stranded loggerhead turtle from Sanibel Island. Just prior to its demise the turtle had ingested a substantial quantity of blue crabs, *Callinectes sapidus*. There were no external injuries and the presence of such recently consumed foodstuff indicates that death occurred very quickly, not as a result of disease, but probably due to asphyxiation because of prolonged forced submergence inside a shrimp net.

Another adult female loggerhead, which stranded on the Sanibel Island bay beach (shore of Pine Island Sound, 200 feet west of the Sanibel Causeway) on Christmas morning, 1975, contained nothing in its digestive tract except the remains of very recently ingested giant seahorses *(Hippocampus erectus)*. Five of these were completely intact.

Fish that are sometimes represented in the digestive systems of loggerheads seldom are chased down and caught by the turtles. Weak or dead fish are certainly consumed. Much of the latter, representing scores of species, originates from fish by-catch that is discarded overboard from commercial fishing vessels, i.e., shrimp boats. In my opinion, vegetation that is occasionally found inside loggerheads during necropsial examination has been ingested secondarily, during the turtle's consumption of animal life.

7
Tags and Tagging

Tagging of marine turtles is not a recent development in the technology that has been devised by researchers to investigate the general life history of the species. As early as 1915, sea turtles were marked in the Dutch West Indies by branding coded letters or numbers on their costal scutes, by drilling through the carapace and attaching silver colored tape, and by drilling a coded series of holes through their marginal scutes. The first serious use of tags by turtle biologists began in 1953 when Tom Harrisson applied an adaptation of the common cattle ear tag to the flippers of green turtles in Sarawak, Malaysia. This early tag was fabricated from Monel metal, a highly corrosion-resistant alloy consisting primarily of nickel and copper. He applied the tags to the foreflippers of green turtles in an effort to identify them later in the course of his work.

Archie Carr experimented with a variety of tag styles during the formation of his pioneering sea turtle studies in Florida in the early fifties. At first, he used a metal carapace tag that connected to the turtle by Monel wires, which passed through holes drilled in the turtle's shell. This design proved to be unsatisfactory because of tag loss. In 1955, he began to utilize the cattle ear tag which his successors continue to apply in their work with *Chelonia mydas* on the Caribbean coast of Costa Rica. Most sea turtle researchers around the world continue to use this tag design in their tagging programs, although in recent years a plastic flipper tag has gained some popularity.

In 1963, I decided to apply the Monel cattle ear tags to loggerheads in an attempt to learn more about the individual turtles within the nesting population that visited Sanibel Island. A year or so before, the Sanibel-Captiva Audubon Society had purchased a series of Monel metal tags for an American alligator study on the island. I used some of the same tags on Sanibel Island *Caretta*.

At 2315 hours on May 26, 1964, George Weymouth and I tagged the first loggerhead turtle on Sanibel. This turtle carried tag 023, and was never observed again. I continued to use the Audubon Society tags on Sanibel's

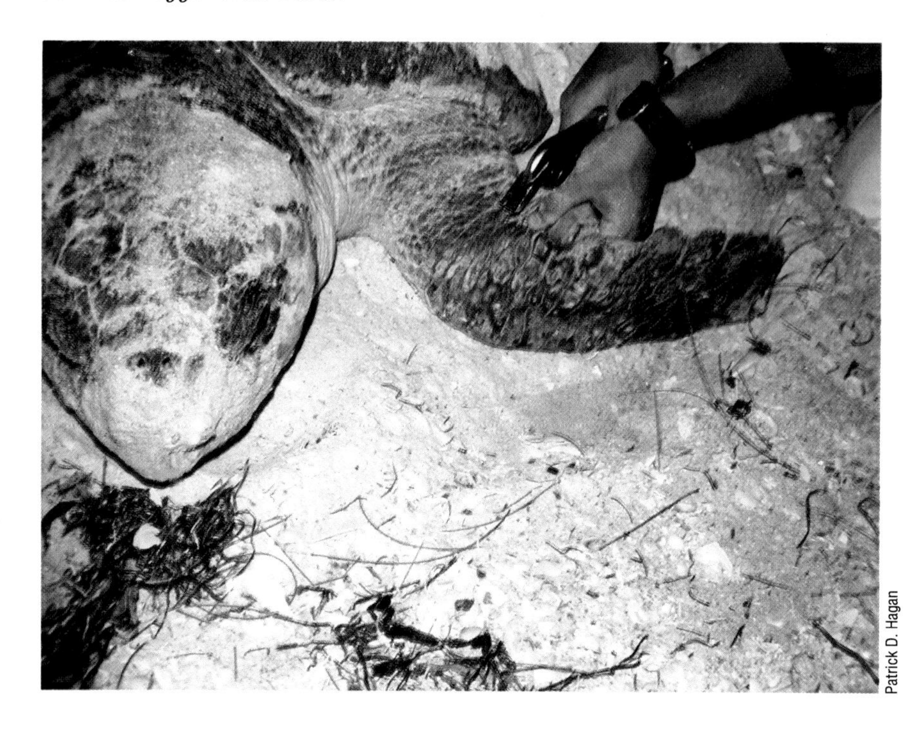

Patrick D. Hagan

Plate 8. A self-piercing Monel flipper tag is inserted into an applicator tool and clamped through the loggerhead's foreflipper. I have consistently applied tags to the left flipper.

loggerhead turtles until the supply was nearly exhausted by the end of turtle season 1967.

Early in 1968, Archie Carr supplied me with a series of his University of Florida tags, and these were used on Sanibel that same year. Caretta Research was formed at the end of the 1968 nesting season. From 1969 until the present, we have used our own series of Monel tags which have borne the letter prefixes, "CR," "SI," and "CL." Since that first Sanibel turtle was so-marked, 1,371 tags have been applied to subregional loggerheads by people associated with Caretta Research, Inc.

Monel tags, in their many modern configurations, are not the perfect tags for sea turtles. They are, however, relatively inexpensive, readily available, and easy to apply. Unfortunately, they are also very easily lost. If improperly crimped by the plier-like applicator with which they are applied, they may fall off. Necrosis of flipper tissue may occur adjacent to the tag causing the tag to fall away. Corrosion of the eyelet of the tag may occur causing the tag to open and be lost. Improved Monel tags with redesigned locking devices are now available. A new alloy, Inconel, a combination of 80 percent nickel, 14 percent chromium, and 6 percent iron shows promise in extending the attachment life of sea turtle tags.

I have continued to use the Monel version because my experience with the product has been good. Inconel tags are very expensive and, at the present, must be purchased in large quantities. A few years ago during a night on the beach monitoring the Sanibel nesting population, Eve Haverfield discovered a female loggerhead which I had tagged sixteen years earlier, with the tag still in great condition!

The Monel metal tags are individually numbered, along with a prefix, i.e., CR 137. This identifies the specific specimen and it is this number, with relevant physical data on the individual loggerhead, that is permanently recorded on field data sheets and ultimately goes into a computer data base program. On the reverse of the numbered side, is a brief message that specifically requests the finder of a tag to notify Caretta Research, Inc.

Some female loggerheads in Southwest Florida occasionally are found with a scar at the approximate site where a tag would have been. To prevent this loss of an individual loggerhead's tag-related data we have improved our tagging methodology. Since 1980 I have been applying a tag of different design on each loggerhead, in addition to the flipper tag. This new tag consists of two 1¼ inch stainless steel fender washers and a ¼ inch stainless steel machine bolt with a nylon insert locking stainless steel nut. With a rechargeable electric drill, a slightly oversized hole is quickly drilled through a rear marginal of the carapace. The bolt, with one washer attached, is pushed through this hole; then the second washer put in place, followed by the nut. The assembly is finally tightened using the proper sized box-end wrenches. This identification device is attached so that the head of the bolt is positioned on the underside of the carapace. In this fashion, any surplus threaded shank which extends beyond the nut is above and does not come into contact with the rear flippers to cause injury to the turtle.

Since development of this tag, I have stamped both washers of the unit with an identical number. Additionally, the washer that is positioned on the topside of the carapace has a three letter prefix, "SAN," stamped forward of the number. In the event of flipper tag loss, the additional backup tagging system that the carapace tag provides maintains continuity of data on the particular turtle. In all cases of flipper tag loss the shell tag provides positive identification and a replacement flipper tag is attached.

Tagging has been restricted, for the most part, to the female gender of the planet's sea turtle populations. And then it only provides information on but a minute portion of the unique life history of these animals. Occasionally, male turtles are tagged on the open ocean by cooperating fishermen, or by scientists who are conducting studies of turtles at sea. In most cases, these turtles are never observed again unless they permanently

Laura Riley

Plate 9. The author and Eve Haverfield installing a carapace tag.

reside in the areas where the studies are being conducted, and they are recaptured, or caught during trawling operations elsewhere, or discovered stranded on some beach. It has been suggested by some researchers that the female loggerhead is migratory and the male is not, but this supposition requires more investigation before it becomes generally accepted by sea turtle biologists.

An experimental technique of marking turtles is being conducted in Virginia where tagged turtles are being injected with tetracycline. This antibiotic is absorbed by the bone and its presence will later be revealed when bone is scanned under ultraviolet light. This concept has been designed to evaluate bone age and its growth rate. Identifiable rings of the affected post-treatment bone are created by injection of the tetracycline. This establishes a known starting point. This methodology is in its infancy and injections have been accomplished on just a few hundred animals. In time, some stranded turtles which were part of this program, may be identified when bone is tested.

Permanently marking hatchling sea turtles has been attempted in many ways. From 1970 to 1975, I surgically notched the rear shell sections

Laura Riley

Plate 10. A carapace tag assembly after application.

of thousands of neonate loggerheads from Sanibel Island. A coded system was designed whereby a specific marginal scute, corresponding to the year of hatching, was removed. The same technique has been employed by sea turtle biologists in other parts of the world, but the methodology has provided little success.

Other techniques which have been tried include: punching holes though marginal scutes, tattooing, applying small plastic marginal piercing tags or wires, attaching disk tags to the tiny rear flippers, surgically implanting minuscule metal balls in flipper tissue, and even inserting radioactive metal particles inside the carapace of hatchlings. One of the more promising new methods that has been developed is a "living tag." This involves the surgical transfer of a small plug section of plastron tissue to the carapace.

The implanted metal balls and radioactive particles may be two of the most promising techniques. Following an established interval of years, selected nesting beaches will be scrutinized by researchers with portable x-ray equipment or a Geiger counter. Nesting females will be exposed to the instrumentation, in the appropriate body areas, in hopes of locating the identifiable implants.

In the early 1970's I utilized a yellow plastic tubular tag, commonly referred to as a "spaghetti tag," on adult loggerheads and also on some of the larger headstarted turtles. Most of these tags were lost during the elapsed time between nesting seasons. The few that remained intact were found to be extremely brittle and their use was discontinued.

Efforts at marking hatchlings have all been attempted in hopes of ascertaining if hatchling sea turtles indeed return to their natal beaches some years later, as adults, to fulfill reproductive needs. Someday a number of green turtles may arrive on some nesting beach, with each of them bearing an unusual area of pigmentation on the same costal scute. This event would settle discussion on the matter for all time. *Chelonia mydas* is the best candidate for the "living tag" approach because of the great contrast in the coloration between the carapace and the plastron.

To encourage the return of tags, I have paid a reward of $5.00 for every tag if supporting information on the turtle's capture is furnished by the captor. I like to think that this has generated interest to return tags in the individuals who may capture or find one of our loggerheads on some foreign or domestic shore. Unfortunately, I have heard rumors of shrimp boat pilothouses which are decorated with strings of sea turtle tags.

To minimize stress to loggerheads during handling for tag application, measurements, and general examination, I avoid turning them upside down. In the case of those animals which are encountered during the egg deposition or clutch concealment stages of the nesting process, most data can be recorded with ease and minor disturbance to the turtle. When loggerheads are encountered during a false crawl or on their way back towards the water, it is difficult for one or two people to immobilize them long enough to take two sets of carapace measurements, get the head measurement, inspect for tags, tag if untagged, and generally scrutinize the animal for injuries, unusual features, etc. It has been customary practice under such circumstances to carefully turn the turtle upside down to prevent escape while routine work is being conducted.

Turning is seldom used on Sanibel Island anymore, unless a turtle is encountered near the surf and must be stopped immediately. Far too much time and money have been expended to allow a specimen to escape unexamined, or untagged, when there is an opportunity to add to the database. In an effort to reduce both the traumatic treatment of turtles as well as the number of persons required to accomplish restraint efforts, a device called a turtle-stopper was created. This was developed in the early 1970's by the late Bob Pond, of Caretta Research, Inc., who used the apparatus successfully on Manasota Key for several years.

The material required for fabrication of a turtle-stopper is minimal in cost. A ten-foot length of one-half inch electrical metal tubing is bent in a semicircle at its midsection with the ends parallel and about ten inches

Dianne Benson

Plate 11. A turtle-stopper, being demonstrated by Ed Phillips, temporarily preventing a postnesting loggerhead from reaching the water.

apart—resembling a giant hairpin. In some models, a sharpened piece of half inch diameter cold-rolled steel is inserted into each end of the turtle-stopper with the pointed ends extended a few inches and welded in place.

To operate the device, the stopper is placed directly over the turtle's neck and against the front of the carapace. The stopper is then firmly forced into the sand by the operator who is situated in front of the turtle. Force is applied against the turtle restraining it long enough to collect measurements, inspect for tags, apply a tag if necessary, and check out the animal for peculiarities.

There are some drawbacks associated with the utilization of a turtle-stopper. It is a cumbersome piece of additional equipment to carry on the beach in anything smaller than an all-terrain-vehicle or a standard size four-wheel drive vehicle. The apparatus works best on a soft beach where the ends can be deeply imbedded in the sand with minimal effort by the restrainer. Soil conditions at Manasota Key were usually excellent for its use. However, the beach soils of Sanibel Island are of different composition, more dense or profusely shelly, making a turtle-stopper sometimes inefficient because the ends will not penetrate deep enough.

8
Reproduction

Warren E. Boutchia

The age at which sexual maturity is attained in wild *Caretta* is unknown. In captivity, with good husbandry, loggerheads can develop all the external sexual features and reach an adult size in about seven years. However, such growth occurs under optimum conditions and is considered to be dissimilar from what occurs in wild populations. Some sea turtle specialists suggest that North American *Caretta* become sexually mature around fifteen years of age. In my opinion, this is a reasonable age range for reproductive capacities to be reached.

Seasonal warming of water in the Gulf of Mexico is one environmental factor that triggers sexual activity in loggerheads. Temperature increases are also commensurate with a higher frequency of sea turtle observations in littoral waters during the non-winter months.

In the mid-seventies I furnished marinas, public boat ramps, tackle shops, etc., with posters requesting reports of sea turtle sightings originating in Southwest Florida waters. Many phone responses were received and numerous personal contacts were made because of this program. The public input provided dates of sea turtle observations, their location, and water depth at the point sightings were made. The majority of the reports originated from sightings that were made offshore between Captiva Island and Cape Romano. Observations were reported from waters which ranged between one hundred yards and thirty-five miles offshore, and where depths varied between ten and fifty feet.

Turtle observations were reported during all months of the year; however, the majority of reports were received during the warmer months, April through October. Loggerheads continued to be regularly observed in the wintertime, in both the open waters of the Gulf of Mexico and in the cooler estuaries and waters behind the barrier islands.

Copulation usually occurs several weeks prior to the start of the nesting season. Water temperatures and length of day are two environmental factors which trigger sexual activity, along with a rise in the levels of the sex hormones estrogen and testosterone. The fertilization processes of sea turtles were discussed in Part One.

During copulation, the male mounts the female, situating his plastron atop her carapace. This is not an easy task and persons who have observed this joining, in rough seas, relate the great difficulty the male encounters in attempting to position himself. Once he is approximately positioned, the enlarged, curved, grasping claws on the front flippers of the male hook over the front of his partner's carapace, and the semiprehensile tail is bent under and forward of the female's vent. The large, horny, terminal scale on the tail of the male serves as an anchoring spur and when the anal areas of the mating turtles are in close proximity, the tail tip is firmly pressed against the female. Once this three-point connection is achieved and penile penetration accomplished, the pair float in unison or are propelled along by the female who continues to have freedom of movement. Usually, the rear flippers of the male are also utilized to hold the female to improve stability during mating. This copulatory embrace, unlike the coital mannerisms of most vertebrates, is often a prolonged affair. I occasionally have sailed near mating *Caretta* while they remained united for up to fifteen minutes—and that was the duration of the union after the pair was first spotted by the on-board observer.

My first observation of a mated pair of loggerheads, which is also the latest record for any season I have available, occurred on May 3, 1957. At the time, I was employed by what today is the National Marine Fisheries Service, and was assigned to the Research Vessel, Kingfish. During a grid run collecting water samples ten miles offshore in the Gulf of Mexico west of Gordon's Pass, Naples, we closely observed a mated pair of *Caretta*. The earliest personal observation of loggerheads engaged in sexual interaction that I have ever recorded occurred on March 19, 1962. This observation was made aerially from a Supercub floatplane, about one mile east of the Sanibel Lighthouse in San Carlos Bay.

Generally, the nesting season for subregional *Caretta* has been described as occurring in May, June, July and August, with nesting activity usually commencing by the second or third week of May and terminating by the fourth week in August. The earliest nest which I have recorded for Sanibel Island occurred on April 27, 1962. The latest nest which I have

Lanny Sherwin

Plate 12. A mated pair of *Caretta* photographed at sea, west of Naples, Collier County, Florida. One of the specialized grasping claws on the male's left flipper is hooked over the front of the female's carapace.

documented was deposited on Sanibel on September 4, 1987. These dates are not a definitive range of a nesting season for *Caretta* in Southwest Florida, for I have heard of loggerheads nesting into mid-September in some parts of the subregion in recent years. Reproducing loggerheads certainly will continue to land on subregional beaches prior to or after the recorded dates for their nesting season on Sanibel Island.

The dates for the first and last nesting landings by loggerhead turtles for the period from 1959 through 1989 are shown in Figure 32. It is interesting to note that during the scope of my record keeping, turtles nested earlier in the sixties and beginning seventies than they did in more recent years. Termination of the respective nesting seasons are relatively consistent for the thirty-one year period.

Like all marine turtles, the loggerhead leaves the water and produces multiple nests during a self-regulated nesting season. This feat relates to egg maturation rates. In the case of *Caretta*, a maximum of seven egg complements have been recorded for a specific turtle. However, not all

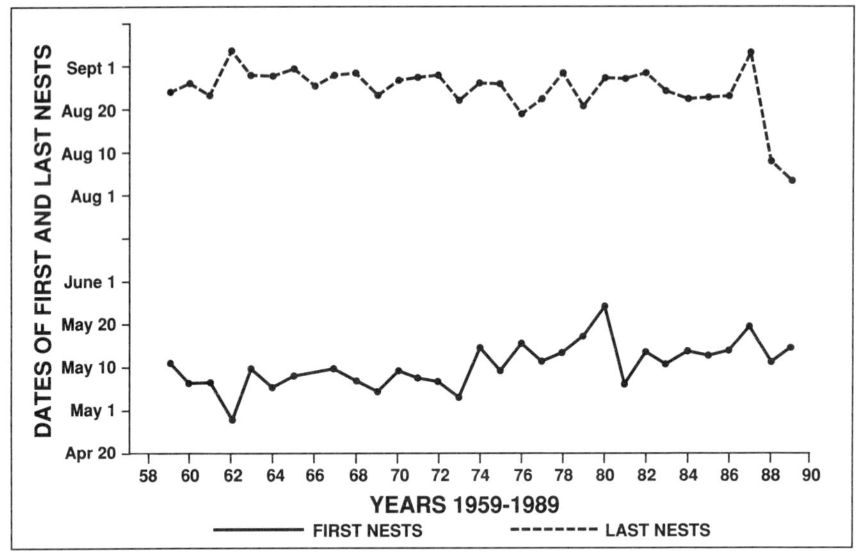

Figure 32. Range of seasonal nesting activity, by year, for Sanibel Island *Caretta*,

female loggerheads nest with this frequency. The average turtle which nests on the coast of the subregion produces about three nests per reproductive season. This information is based on observations of tagged animals during their repeat visits. The time span between nesting seasons appears to be self-regulated, although the majority of subregional loggerheads nest every other year. In other words, part of the population nests in even years and the other portion during odd years. I do not mean to suggest that this frequency becomes routine and is maintained throughout an individual's reproductive life. It is known that turtles can alter this cycle and convert from biennial to annual and even extend to a triennial nesting schedule.

The time span between the in-season nesting emergences is known as the internesting interval, and this is variable throughout the loggerhead turtle's nesting range. Over the years, multiple nestings by tagged turtles have been examined in the subregion to determine the periodicity between deposition of these multiple clutches. The average internesting interval for the population of *Caretta* nesting on Sanibel Island is about eleven days. This is shorter than the fourteen day interval generally stated as normal, or average, for loggerheads; however, shorter, or longer, internesting intervals are known for the species. Some researchers have reported intervals of up to twenty-eight days and have attributed such extended intervals to the intrusion of colder ocean water in the vicinity of the nesting beaches. However, some loggerheads make unexplained shifts in their nesting beaches, and the possibility exists that nestings were not delayed because of temperature. The animals may have made an unobserved intermediate nesting visit to a nearby section of beach.

Such environmental parameters probably can decrease or increase the maturation rate of egg clutches being processed in the reproductive systems of females. It is believed that most gravid loggerheads remain offshore from their selected nesting beach and that this behavior may account for the site fidelity common for later nests. It is amazing just how closely female loggerheads land to their previous nest sites during subsequent nesting landings. This occurs after nearly a two week interval, during which time they must continually compensate for variable offshore winds, tides and currents.

Egg-laden loggerhead turtles intent on egg deposition usually begin their inshore swim before, or just after dark, depending on an individual timetable. Arriving near the water's edge in darkness, they may be assisted landward by high surf or they must use their own slow, cumbersome locomotion to come to rest near the surfline. Thus begins the first stage of a procreative event that has become standardized because of its simple success through the ages. The nesting behavior of *Caretta* is extremely functional and for all intents and purposes is not divergent between individuals.

In some parts of the world, sea turtle nesting activity has been shown to coincide with certain tidal cycles. Many people working with sea turtles have reported that nesting activity, landing on the beach, etc., occurs during flood tide. This may be a reasonable consideration where the tidal amplitude is great, but I have found no correlation between tidal stages and nesting activity. In Southwest Florida, the standard tidal deviation is under two feet, unless weather events play a role in the elevation of the tide. Subregional loggerheads emerge from the surf and complete nesting regardless if tidal conditions are flooding, ebbing, high or low.

Awash in the surf, the loggerhead female soon begins stage two of her journey. The ascent of the beach may be laborious and time consuming, depending on the profile and soil consistency. The individual may pause, with head elevated as if visually examining the beach-scape, or appear to alternately probe the beach surface with her snout. If head weight is the factor which causes sea turtles to seem to force their head into the sand, then the massive head of the loggerhead certainly fits this possibility. Visual signals of the beachfront being scrutinized, assuming olfactory cues are of no real consequence, may help the individual recall earlier memorized landmarks and identify the beach as the preferred location from which to leave the water and nest.

With alternating manipulation of her flippers, the loggerhead begins to move ahead and slowly drags her heavy body across the wet beach. Periodically she may pause to rest, again snout-probe the sand, or raise her head and peer about the landscape. The ascent trek may be short, ending just above the spring tide line. On wide beaches it may be long, ending a hundred or more feet from the water's edge. Some individuals, once they reach soft dry sand, bulldoze their way along and leave shallow trenches which may be yards long.

Many factors influence the site selection process, the third behavioral mode of a successful nesting venture. The beach traverse by loggerheads may result in a non-nesting emergence—a false crawl. When all conditions of the site meet the inherent requirements of the individual turtle, the activity moves on to the fourth behavioral characteristic—site preparation or clearing.

Members of the marine turtle genera *Chelonia* and *Dermochelys* excavate depressions or body pits in the beach surface. This action in itself usually accomplishes site preparation. *Caretta* makes a minimal attempt at body pit excavation. When the site has been chosen, sweeps of the foreflippers broadcasts soil to expedite its removal. Simultaneous side to side movement of the shell also helps to remove the dry topsoil, bringing it into reach of the arching front flippers and the smaller rear flippers which are also throwing sand aside. This action moves the turtle a few inches lower than the beach surface, creating a minor body pit. The resultant depression is insignificant compared to those created by the green and leatherback turtles. On some beaches with steep profiles, the front flipper sweeps of a loggerhead may dig deeply into the beach and give the appearance that the site has been developed considerably more than it actually has. The removal of dry, loose sand from beneath and immediately adjacent to the turtle makes excavation of an egg chamber less arduous. The displacement of the top sand eliminates material that would otherwise slough into the egg cavity making its formation difficult.

The fifth step is the formation of the unique egg chamber. The excavation can be very time-consuming, depending on the compaction of the beach soil and the presence of solid, buried objects, such as large sea shells, roots, or logs which may impede the digging process. The digging starts with one rear flipper while the front flippers are held closely to the shell and securely imbedded in the sand. After several probes are made into the soil to loosen it, the flipper is curled inward to lift out nearly a cupful of material, and deposit this to the outside perimeter of the soon-to-be egg chamber. Next, the opposite flipper is brought into play; the appendages are utilized alternately through the rest of the digging activity. Figure 33, on page 63, is an excellent pictorial sequence of the egg chamber excavation.

In drawing A, a diagrammatic loggerhead turtle removes soil with the left rear flipper as the right flipper retains the sand which earlier was removed from the right side of the cavity in the previous sand-probing and

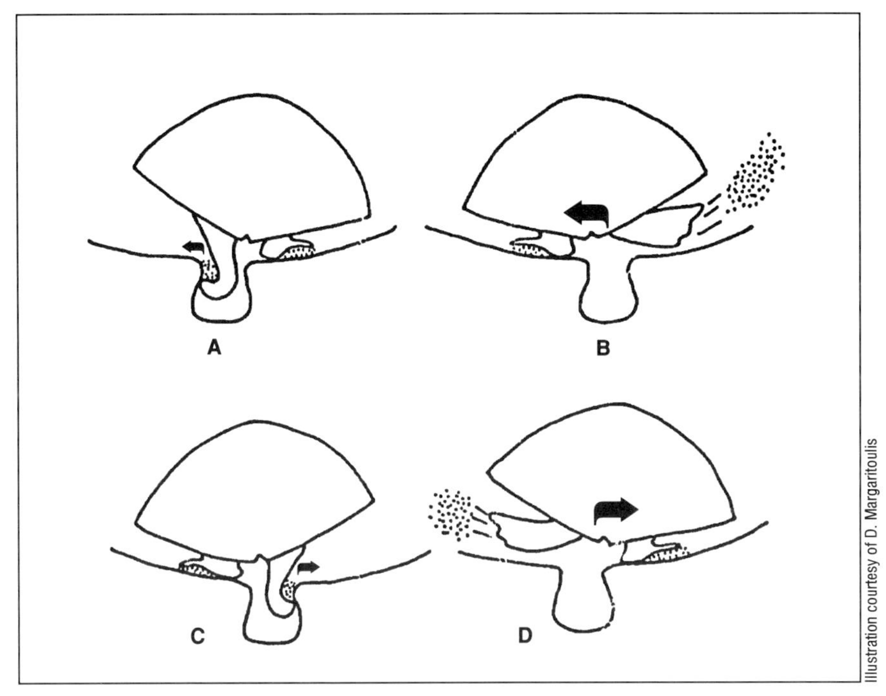

Figure 33. The four-part nest excavation cycle utilized by nesting *Caretta.*

material-lifting segment of the excavation cycle. In B, the sand removed in phase A is dropped on the left side of the nest opening and the turtle shifts her weight to the left and holds the loose soil in place. Simultaneously, the right flipper broadcasts, with a quick flick of the flipper, the sand that had been removed during the previous stage of the four-part cycle. Drawing C illustrates the turtle's action as sand is removed from the right side of the cavity with the right flipper and dropped on the right side. In D, the loggerhead repositions her weight to the right and throws the soil aside with the left flipper. This serialized description of the egg chamber excavation by a loggerhead turtle is based upon the very fine work of D. Margaritoulis which he is conducting on the *Caretta* population of Zakynthos, Greece. He kindly provided permission to incorporate Figure 33 into my work.

The repetitious digging continues until the turtle can no longer bring forth sand from the chamber. As digging movements probe into the nest hole and the flippers reach their maximum extension, they scrape the lower sides of the cavity in a semicircular sweeping motion in order to dislodge and remove the final remnants of loose soil. This creates a chamber that is flask-shaped in cross-section with a round shaft of greater diameter at the

Laura Riley

Plate 13. An egg cavity filled to near capacity. The pink-colored ovipositor, forward of the tail tip, is still extended.

bottom than at the top. Egg cavity width and depth are variable and depend on the size of the female. An average nest cavity has a top opening diameter of around seven inches, a bottom diameter of about nine inches, and a depth, from beach grade to the bottom, of approximately nineteen inches.

The next stage of nesting, the sixth, is the actual deposition of the nearly perfectly round, pliable-shelled, mucous-coated, off-white eggs. When the rear flippers are fully extended and their lateral scraping sweeps can no longer reach sand at the bottom of the chamber, they are withdrawn and placed above and to each side of the chamber opening. With the rear flippers positioned well apart on each side of the freshly dug hole, the egg laying process can easily be seen by an observer peering under the carapace. The turtle will raise and lower the rear of her shell several times and slowly the pink tissue of the ovipositor extends from the cloaca. This extension serves as a tube through which the eggs pass and drop into the egg chamber.

The trailing edges of the rear flippers are curled upward and then relaxed as the female exerts muscular force to expel the eggs. Slowly, one by one, the eggs drop. Each time an egg, or eggs, pass through the oviductal opening, they are preceded by this regulated flexing and the relaxing of the rear flippers. Between the expulsion of eggs, a heavy mucous fluid streams from the body opening and each egg is dampened by this substance. This liquid serves as a lubricant as the eggs travel along and through the oviduct. It may also supply important early moisture for the initial development of the eggs. There is some evidence, too, that the clear fluid serves as an antibacterial agent which provides a level of protection for the eggs.

Once oviposition begins, the nesting turtle is almost completely oblivious to what is going on about her. The expelling of eggs becomes very mechanical—flippers flex and eggs drop. The eggs usually are deposited singularly at first, but soon begin to drop in groups of two's, three's, and four's and continue in such numbers until egg-laying ceases. When the egg chamber is filled to the extent that the eggs on top of the clutch contact the tissue of the ovipositor, deposition ends; however, this is not always true for I have observed a few clutches which overflowed the chamber and eggs were broken by the female during covering behavior.

Following a brief pause, stage seven of the nesting event starts as the dexterous rear flippers begin to scoop soil over the eggs. Following the initial covering of the egg complement, each flipperful of soil is kneaded in position after it has been placed over the location. The full weight of the loggerhead's posterior is utilized to tightly compact the soil over the eggs during this stage. On rare occasions I have heard, as a turtle is energetically transferring her body weight from side to side, an audible thumping sound as the lateral and rear margins of the carapace strike the underlying soil. After an egg clutch is nestled in the cavity, between five to seven inches of soil are placed over them by the female during the covering task.

The next stage, eight, involves general site concealment, or as it has sometimes been called, nest site camouflage. With broad, full sweeps of the front flippers, the loggerhead broadcasts a substantial volume of sand behind her, nearly covering her carapace in the process. She may inch ahead as she tosses quantities of soil over her carapace and the nest site. After another short rest period the turtle turns, slowly, and begins the return trip to the surf, and beyond. Usually, the last stage of the nesting process, the beach descent, is relatively uneventful. Following the turn from the egg site, the route is normally direct to the surf. However, the move back to the water may be accompanied by a rest pause or two for what may be a visual reconnoiter of the terrain and selection of direction. The turtle normally parallels the ascent track with a few feet separating them, although she may travel directly over the up crawl and create a track, that on casual examination, appears to be one-way. Occasionally, other factors can influence the return trip and some of these will be discussed.

The time frame for landing on the beach, ascending the beach face, site

Rickie Cooper

Plate 14. The final stage of nesting before turning and descending the beach, consists of quantities of sand being thrown back by the front flippers.

selection, site preparation, egg cavity excavation, egg deposition, clutch and nest site concealment and return to the water generally takes between one and two hours. Many factors influence the length of time that a female loggerhead turtle is out of her element: width and incline of beach, soil consistency, vegetation present at the selected nesting site, physical impairments, and clutch size.

Copious, thickened, tear fluid is secreted from the loggerhead's eyes when females are ashore for nesting. Over the years, the tears of sea turtles have been surrounded with speculative reasoning as to their purpose. Biologically, their function is osmoregulatory, although they do serve other purposes too. Early casual opinions which have become part of folklore, insist that tears are produced as a result of the physical pain produced during egg deposition, or of the great sadness generated within the female turtle because of fear for her offspring's fate. Both fables have been dismissed by modern sea turtle biologists. Another, more reasonable, explanation of tear secretion is that the tears help to flush wayward sand particles from the eyes during the nesting process.

The lacrymal glands that handle the important task of osmoregulation in *Caretta* are located on each side of the head in the eye orbit. The individual gland is large, over five centimeters long in adults, and is divided into approximately one hundred lobes which are surrounded by blood vessels and connective tissue. The rusty colored glands open to the posterior corner of each eye through a short, wide duct.

During the summer of 1976, I measured the salinity of lacrymal secretions from several nesting loggerheads on Sanibel Island. The salinity of typical water samples which were taken from the Gulf surf during this short-term study, averaged 33.4 parts per thousand (ppt) of dissolved salt. A salinity refractometer was utilized to determine sodium chloride levels. The instrument was cleaned and rinsed with fresh water before each use and samples of lacrymal secretions were taken from dry loggerheads following egg deposition. Only clear sand-free fluid was examined. Salinities recorded ranged between 24.4 ppt and 47.5 ppt, with a mean of 37.8 ppt, or slightly higher than the normal Gulf water in which these animals live. This brief study clearly indicated that tear secretion levels are hypersaline; that is, condensed salts were being eliminated through the lacrymal glands.

In 1973, I encountered a loggerhead as she attempted to deposit her eggs after having excavated the egg chamber in typical fashion. As she began to extend the ovipositor through the vent, I was surprised to see that the tissue was covered with hundreds of nearly inch long worm-like organisms. The opening of the ovipositor was completely sealed by these animals, and a hard calcified deposit prevented complete distention of the ovipositor. I examined the individual and discovered a general infestation of the parasites at all other body openings—the eyes, nostrils, and on each side at the rear of the turtle's mouth. I attempted to collect some of these parasites by detaching them from their host with a stick but was unsuccessful. Whenever I would touch them with my fingers, trying to pull some loose, they would withdraw into themselves and I could not get a grip to pull one loose.

During this time, the turtle was straining to expel her eggs which could not pass because of the combination of parasitic biomass and the adhesion of the hardened material. Ed Phillips and I spent nearly an hour with this poor animal, alternately trying to collect a sample of the parasite and clearing the oviductal opening. If we would have had forceps or some other tool in the vehicle, it probably would have been possible to remove a sample.

Finally, my probing must have proved to be too uncomfortable for the turtle and she suddenly left the site and returned to the water. I remain convinced that there was no way this loggerhead could have performed oviposition. This female's plight and inability to oviposit raises a little understood aspect of chelonian reproductive biology. It is known that some turtles are capable of retaining mature eggs internally for weeks should environmental, habitat, or other factors not be conducive for timely depo-

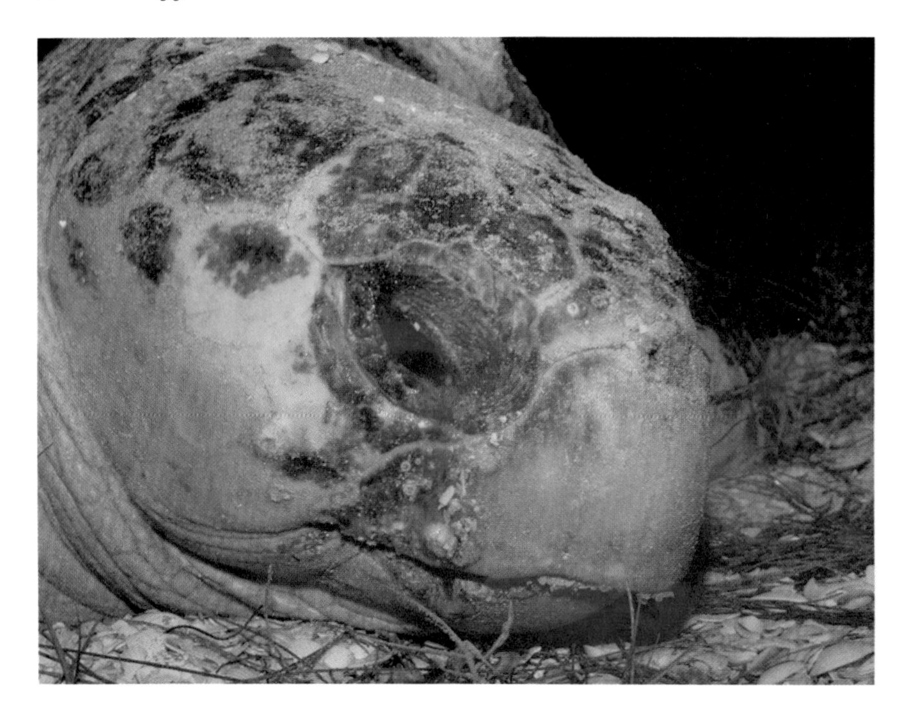

Plate 15. Lateral view of the head of an adult female loggerhead. The liquid being secreted by the lacrymal glands is visible beneath the eye.

sition. Retention, or developmental arrest, cannot be sustained indefinitely and normal physical development of embryos will not proceed if the period of retention is prolonged. Since the individual discussed could not successfully accomplish oviposition of the eggs she was carrying, they were probably eventually resorbed. Egg resorption is known to occur in reptiles but the mechanics are not fully understood. Premature eggs may be resorbed rather quickly when they are merely ovarian follicles; however, the time framework required for complete resorption of mature eggs is unknown.

It seems to me that the parasite infestation of this loggerhead was of recent colonization. Not only could eggs not pass from the oviducts, but I also am doubtful that the penis of a male loggerhead could have penetrated this female to accomplish fertilization. Undoubtedly, all evidence indicates that spermatozoa had reached the appropriate reproductive organs and this turtle was gravid. The actual entry of semen may have occurred at an unknown time before the aggressive parasites had blocked her cloacal opening to such a remarkable degree.

Years later, I described these organisms to Erik Martin, who is with

Applied Biology, Inc., of Jensen Beach, Florida, and he identified them as *Ozobranchus margoi.* This is a species of marine leech that is known to occur on sea turtles and dolphins in many parts of the world. It is quite common along the eastern seaboard of the United States, and about 20 percent of *Caretta* in the vicinity of Hutchinson Island, in St. Lucie and Martin Counties, Florida, are infested to varying degrees. The leech is not as common on loggerheads that nest in the subregion.

In the early years, I recorded the direction that postnesting loggerheads selected on their return to the water. Some individuals made a clockwise crawl while others went in the opposite, or counterclockwise, direction. Field data sheets, from 1975, reveal that thirty-one observed loggerheads made clockwise crawls and thirty-seven made counterclockwise crawls on Manasota Key. On Sanibel Island, all turtle crawls were routinely inspected for direction until the mid-seventies. It is interesting that there is a fifty-fifty distribution of direction, or turn preference, in that nesting population too. When the return crawl direction preferences are reviewed, it becomes clear that a specific turn direction selected after an earlier nesting is not sustained by the turtle in later nestings. An individual may move counterclockwise after completing her first nest of the season and clockwise following a second nesting, and then switch back and forth during any subsequent nesting visit.

As female loggerheads leave the water, they generally travel at a right angle from the surf's edge. This route may not be exactly perpendicular and may angle somewhat to either side of a true ninety degree bearing. When the turtle arrives at the selected area where she will begin site preparation, and later egg deposition, her head almost always faces inland. In fact, a very limited number of loggerhead turtles which I have observed over the years, have varied from this norm. Only half a dozen times during my long career with loggerhead turtles have I had the opportunity to observe a female nest while facing the water. When this rare modification of behavior occurred the animal had apparently false crawled, was enroute to the water, but suddenly stopped while in the nesting zone and proceeded with the multifaceted nesting process.

Loggerhead turtles are usually nocturnal visitors to their nesting beaches. Diurnal nesting by *Caretta* occurs infrequently. Over the years, I have received reports of turtles ashore near mid-day during the nesting season. On those occasions, I would race to witness the rare event and, undoubtedly, with the exception of one instance, the turtles were always gone when I arrived on the scene! Years ago, after some long night on the beach, I sometimes have encountered nesting loggerheads after sunrise. In comparison, reports of loggerheads ashore for egg deposition as early in the evening as 1830 hours have not been uncommon. I have never classified such turtles to be diurnal nesters, but consider them to be animals which

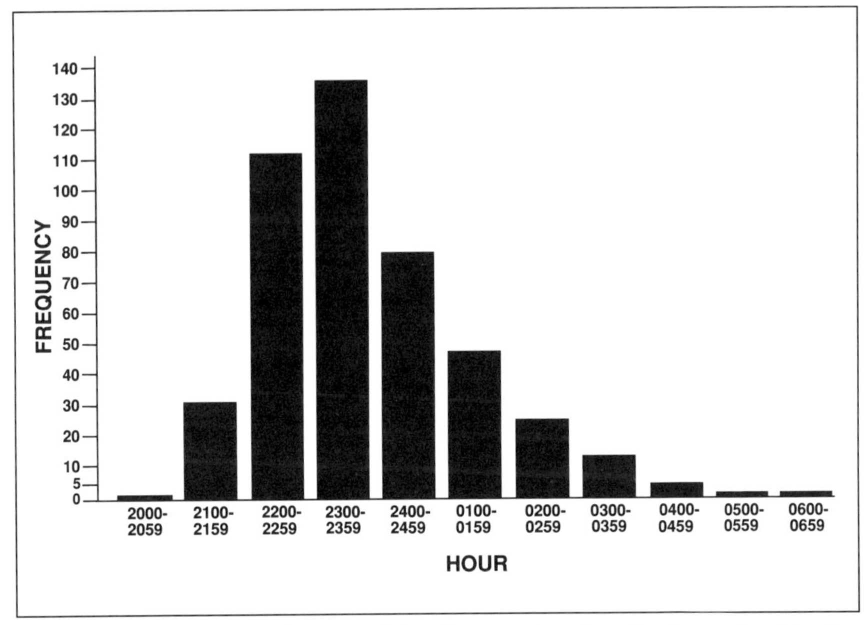

Figure 34. Distribution of nocturnal time frequencies of nesting loggerhead turtles observed on Sanibel Island. Based on tagged and untagged turtles, 1964-1980. Eastern Daylight Saving Time.

landed on the beach during predawn darkness and remained on the beach after daylight. Those loggerheads which have been discovered at dusk are likely animals who made good time during their inshore swim, and then arrived at the nesting beach early in the evening before sunset.

Two mid-day nestings have occurred for which I can provide documentation. On June 6, 1973, at 1140 hours, Paul Zajicek and his coworkers at my Cape Romano study beach, had a loggerhead turtle (CR 441) interrupt their lunch as she emerged from the surf and ambled into their campsite. There, she went about the chore of site selection, nest cavity excavation, egg deposition, clutch concealment and returned to the Gulf—all normal behavior, but under a scorching summer sun and in the presence of an astonished turtle conservation team. Interestingly, this same loggerhead was encountered three additional times that summer. On June 27, this turtle was found at 2145 hours (109 eggs), on July 7 at 2350 hours (123 eggs) and on July 19 at 2220 hours (130 eggs). The daytime clutch produced by CR 441 on June 6 totalled 99 eggs.

More recently, a loggerhead (CR 5075, SAN 118) emerged from the water at 1130 hours on July 28, 1984, near a Sanibel Island resort, passed through a group of awed tourists who were strolling along the beach, and

Plate 16. A typical loggerhead turtle nesting crawl on the Sanibel Island beach.

nested successfully. I was notified by the resort owner and was fortunate enough to arrive before the turtle had completed egg laying. When she was through, I measured and tagged her. I also had my first opportunity to get a different series of loggerhead turtle nesting photographs—without flash equipment.

It has been suggested that diurnal nesting by loggerheads may be related to the timing of an animal's inshore passage to the proximity of the time of high tide. High tide is thought by some biologists to assist land-approaching turtles pass safely over reefs or near-shore shoals. If this rationale were acceptable, would it not stand to reason that most sea turtle species would not be nocturnal nesters, but would always take advantage of high water? However, as discussed earlier, this is not the case. I have found no clear correlation to suggest that nesting timing for the subregional *Caretta* population is subject to tidal considerations.

The tide-related timing of the two mid-day nesting loggerheads which were discussed, provides little credibility to support the notion that high tide plays an important role in the frequency of diurnal nesting events. The Cape Romano loggerhead travelled to the beach and came ashore at 1140 hours on June 6, 1973. On that date, low tide near Cape Romano occurred at approximately 1148 hours. Loggerhead turtle CR 441 would have had to make a shoreward swim during an ebbing tide, landing just minutes before the tide changed to the flow mode. Loggerhead turtle CR 5057 casually ventured ashore on Sanibel Island at 1130 hours, just forty-five minutes prior to high tide on July 28, 1984. This individual may have been assisted to the beach with the interaction of flow tide, and also was helped in terms of the amount of physical energy she had to expend, because the higher the tide, the shorter the distance to the beach nesting zone.

Diurnal nesting by *Caretta caretta* in modern times may be a recessive trait genetically. It has been suggested by some marine turtle biologists that at one time, perhaps eons ago, all sea turtles made their reproductive visits to the land during daylight. It is further opined that early sea turtles on land during daylight, were preyed on by large diurnal terrestrial carnivores. The populations, over time, changed their reproductive behavioral scheme because those turtles with the daytime nesting trait were removed by predators, while those with genetic coding for nighttime nesting survived. Nocturnal nesters now dominate the sea turtle group with the exception of the genus *Lepidochelys,* for ridleys continue to visit their nesting beaches, in groups, during daylight. It may eventually be shown that an occasional sea turtle, other than one of the ridleys which carries the now recessive genetic characters for daytime nesting, makes an alternate nesting visit during daylight.

9
Non-nesting Emergences

Sometimes a female loggerhead turtle leaves the water for what is intended to be a nesting attempt, and then suddenly returns to the sea without selecting a nesting site. These crawls may leave short horseshoe shaped tracks, or they may be long and reach or surmount the pioneer beach vegetation zone before the turtle turned around. There are a variety of reasons for false crawls, as they are commonly called, and I will discuss their causes based on my observations. Sixty percent of all the emergences by *Caretta* on the shores of the subregion are false crawls. I have grouped those factors which contribute to non-nesting emergences into two classes. The first of these is natural or environmental causes and the other is direct interaction with man, or as a consequence of alteration of the ecological parameters.

Typical subregional beach profile types are shown as diagrammatic cross sections in Figure 35. Each major profile type is identified and described in this section. Preferred nesting beaches generally have an inclined profile between the water's edge and the upper dune vegetation zone (Beach profile A). Short beaches, that is, those with a minimal traversal distance required, seem optimum. A loggerhead moves across the dry beach with laborious difficulty and the shorter the route of travel to a nest site, and back, the better. The distance between the mean high tide line and the pioneer vegetation of the frontal beach is extremely variable. Major differences in the width of the beaches which fit within the general categories that I use are common. One Type A beach on Sanibel Island may be considerably wider than a Type A beach near Naples.

Beaches on the Southwest Florida barrier island chain which have a profile consisting of frontal berm, followed by a depressed berm, and then a prevegetation berm, are beach types where false crawls occur very commonly (Beach profile B). Turtles seem to pause when their intended route takes them into a frontal beach or berm depression. Most loggerheads which I have observed over the years, have turned about at the specific point where the berm depression incline begins and headed back to the surf.

Beach profiles are shaped by many factors. Often, just the right combination of tide and wind direction will produce a section of high escarpment in the frontal dune at the spring tide line that is insurmountable for a loggerhead (Beach profile C). Usually, under these conditions, the turtle will attempt to climb the abrupt rise and, if unable to do so, will parallel the escarpment for a short distance, make additional fruitless attempts, then make a turn and false crawl. Low escarpments, those up to two feet in height, are generally scaled by loggerheads. It is the three or four foot high erosion ledge that almost always prohibits their passage. On rare occasions, I have been surprised to find C profile beaches with escarpments between four and five feet high, where a turtle had the fortitude to undermine the barrier and, in her determination, actually construct a steep ramp which she was able to negotiate with what must have been great physical effort, and successfully nest. Once I encountered a loggerhead who appeared to be desperately trying to climb a two foot high erosion ledge and was having difficulty. I interceded and every time she would actively try to climb the escarpment, I would lift her by the rear of her carapace and push ahead in tune with her efforts. After about four exhausting attempts for both of us, the turtle and I managed to get her where she wanted to go, and where she wanted to nest.

Similar, but insurmountable, circumstances occur should a bulkhead or seawall exist in lieu of the natural erosion-produced barriers (Beach profile D). Many beachfront communities along the coast of the subregion, which were developed prior to State or local regulations prohibiting construction out on the beach, have buildings standing on the foredune systems which are fronted by concrete erosion protection devices.

Severely eroded beaches are extremely narrow with little dry open beach above the high tide line. Trees are undercut and fall into the surf as progressive erosion forces encroach on the upland vegetated system (Beach profile E). This beach type is quite similar to profile C, but in no way could a loggerhead turtle, not even a "ramp builder," reach the top for nesting. This is not a disadvantage, since severely eroding beaches offer little stability and chances for full-term egg development are minimal. More than likely eggs which are not located far enough upland in a profile C beach system would tumble into the surf with the first storm tide.

Many other factors contribute to the causes that result in false crawls. Should a loggerhead find her path obstructed during her ascent by a large fallen tree on an eroding beach, or even a tree trunk placed on the beach by the tide, the turtle will often abort the site. Some individuals try to climb over, or skirt, the obstacle; however, this maneuvering is not always attempted. Clumps of thick sea oats or other dense beach vegetation often prevent excavation of an egg chamber and this too results in a false crawl. One of the major impediments to successful sea turtle nesting on subregional beaches is the presence of the Australian pine, *(Casuarina equisetifolia)*.

On the Gulf beaches of Southwest Florida, *Casuarina* is a constant

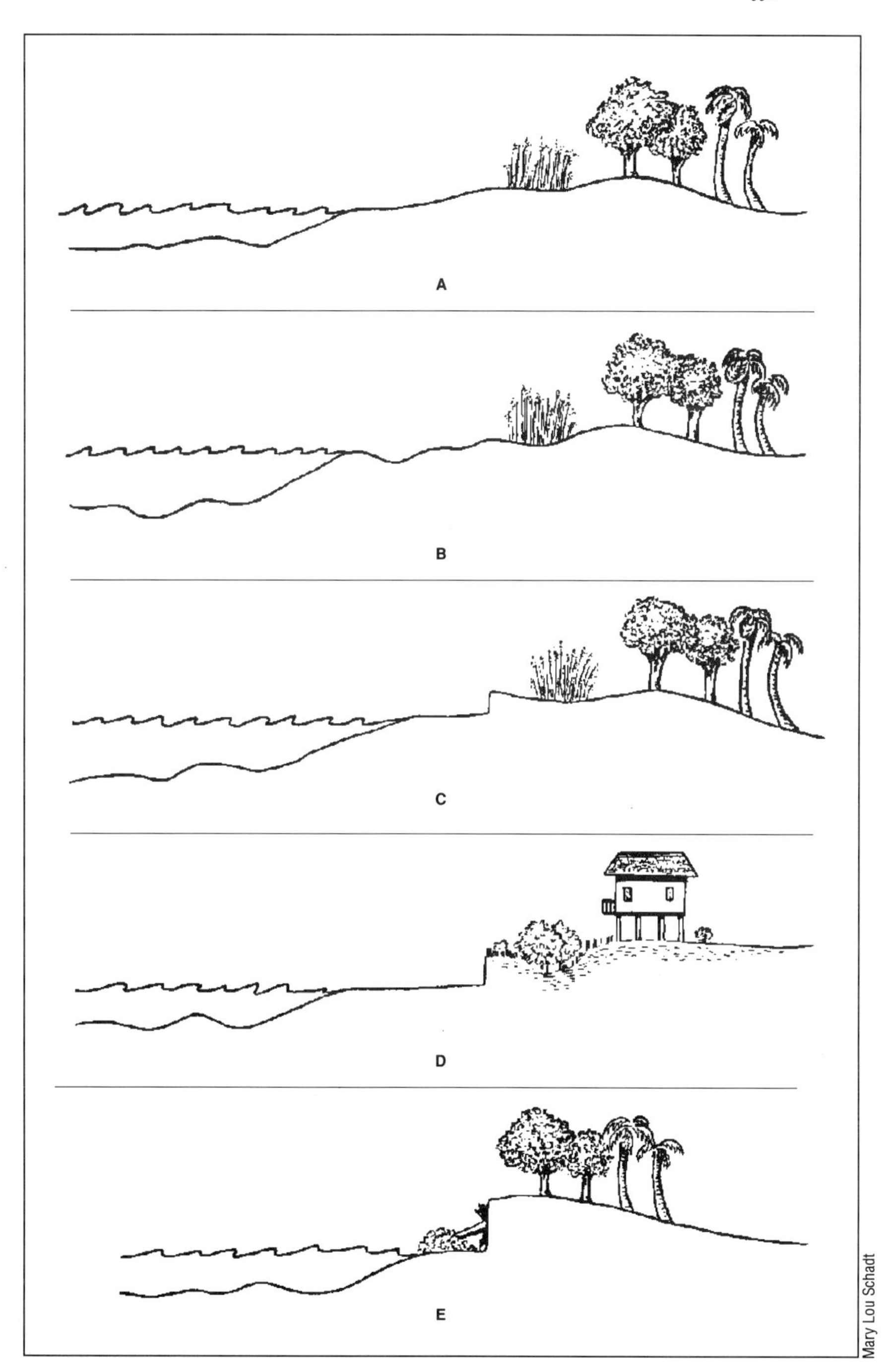

Figure 35. Common types of subregional beach profiles.

invader of the beach ecosystem. These trees continually encroach on the upland beach and rapidly out-compete the more desirable native beach plants. These monotypic stands of *Casuarina* are created as a result of the damaging effects of storm tides that rip away, or cover with a layer of fresh sand and shell, the existing beach vegetation. The disturbed soil which is left in the wake of storm events is a prime nursery for the germination of *Casuarina* seeds. On beaches where this exotic plant is the dominant species, the trees grow in graduated canopy heights. Closest to the water, seedlings grow in profuse bands that parallel the spring tide line. Between this beach zone and the upper beach are saplings which are well established. In turn, the upper part of the beach is covered with mature seed-producing trees. Beachfront *Casuarinas* may attain a height of thirty feet in five or six years.

The multitudinous roots of this tree are very close to the beach surface, and they create a buried web-like network only a few inches beneath the sand. A sea turtle, finding herself located near one of these almost impenetrable stands, is indeed hard-pressed to excavate an egg chamber through the myriad root system and ultimately aborts the task in a false crawl.

On the higher energy beaches, where seasonal erosion or storms eat away the low profiled foredune, *Casuarina* tumble into the surf. The exposed roots and leveled trees present a barrier that, in many cases, eliminate any hope of a loggerhead turtle getting much above the wash of the surf. This is a very common predicament on many of the subregion's barrier islands. Some of Bowman's Beach, a three mile section of remote beach at the northern end of Sanibel Island, is virtually useless for nesting by loggerheads because of the presence of the noxious Australian pine.

I have made it a personal mission to eliminate *Casuarina* seedlings and saplings from the foredune and other areas typically included in the loggerhead turtle nesting zone on the Sanibel beach. For decades now, I have encouraged my coworkers and others who participate in the summer sea turtle patrols, to pull up these small trees whenever there is an opportunity. It has been customary while we are stationed around a turtle and awaiting completion of egg deposition, to police the vicinity and enhance the beach environment by physically pulling out every single *Casuarina* in view that is manageable.

Another natural aspect which often will result in an unsuccessful nesting visit to the land is the composition of the beach soils. This phenomenon is very common on the western section of Sanibel Island because of the island's biogenic formation. Frequently, a loggerhead may emerge at a location where the upper beach nesting zone consists of spotty layers of coarse shell fragments or nothing except complete halves of small bivalves, i.e., ark shells. Attempts at egg cavity excavations in such situations are fruitless because of the unstable consistency of the shell material which

Paul Zajicek

Plate 17. The Gulf of Mexico beach of Kice Island near Cape Romano, Collier County, Florida. Erosion of this beach has resulted in exposed mangrove stumps which can interfere with nesting loggerheads. In this type of situation, native vegetation can be as detrimental to nesting as introduced plants. Paul Zajicek, and his coworkers, once rescued a loggerhead turtle, near this site, that had become hopelessly trapped in such a myriad of obstacles. Had they not discovered the turtle, it would have succumbed.

continuously sloughs into an ever-widening depression. I have observed a number of individual loggerheads abort and false crawl when impeded by such conditions. Sometimes, determined individuals will move ahead a few feet and try again a second, third, or even a fourth time. Often, just a short distance away from previous cavity attempts, the stratified composition of the soil will be more favorable and those more tenacious individuals will successfully nest. On a few occasions, I have also observed nesting loggerheads that have struck a buried object, such as a piece of wood or a large univalve, i.e, a very large conch or whelk. When the turtles were stymied in this manner during nest construction, they invariably false crawled, making no noticeable effort to move higher on the beach to try again.

The presence of animals on the beach will frighten the turtle or otherwise discourage completion of normal nesting behavior. Egg seeking raccoons are the primary culprits. Dogs, too, will quickly dissuade a loggerhead that is trying to fulfill her reproductive needs. I can clearly recall back in the summer of 1961, watching a band of five raccoons dig away the sand behind a nesting loggerhead and deftly remove each and every egg as it dropped into the sandy chamber—until I drove the predators away.

Human nighttime activity on subregional beaches is also a factor closely associated with the occurrence of false crawls. Uninformed tourists and residents who happen on an emerging or nesting loggerhead without hesitation, will turn on flashlights and even poke at or step on the animal. If the behavioral mode of the turtle is not locked into egg deposition, she becomes alarmed, discontinues her actions, and false crawls. This is a common occurrence where high levels of after dark public use take place. The Gulf-front resort areas of Sanibel Island are crowded with shellers, at low tide and regardless of the hour. Sanibel's reputation as the shelling Mecca of the northern hemisphere puts additional stress on the island's nesting population of loggerheads. Telltale track patterns of non-nesting emergences in these high use areas are common.

Figure 36 consists of a variety of diagrammatic loggerhead turtle crawl configurations. In crawl A the turtle emerged from the surf and confronted a bulkhead. Unable to ascend the beach and reach the nesting zone, the frustrated turtle returned to the water. Crawl schematic B results when an individual reaches the beach berm, but turns about when the berm drops into a dry water-paralleling and relatively deep depression prior to the rise in elevation of the foredune. The track shown in D is simply a horseshoe-shaped crawl that may not surmount the high tide line or be some distance seaward of the optimum nesting zone. In D crawls the turtle has turned, for no apparent reason, and returned to the sea. Many crawls of the D type never get above the wet beach and are obliterated by the next high tide. The peaked formation of an E type crawl usually indicates the turtle was alarmed and abruptly returned to the water to escape some threat.

Crawls which do terminate in a successful nest, and are included in Figure 36, will be reviewed here. In crawl diagram C, a turtle has left the water and crawled headlong into a moderately steep erosion escarpment. The animal attempted to surmount the ledge, was unable to do so, and moved to the left where another try brought her to the high beach for normal nesting. Diagrams F and G are normal nesting crawls, where there have been no obstacles in the individual's path. Turtle F reached the high beach and deposited her clutch in, or very close to, the beach vegetation. Loggerhead G moved a short distance up the beach just above the spring tide line, and deposited her eggs out on the open beach about equidistant from the high tide and vegetation lines. Turtle H made a standard emergence, moved to mid-beach, and nested; then she was distracted by the

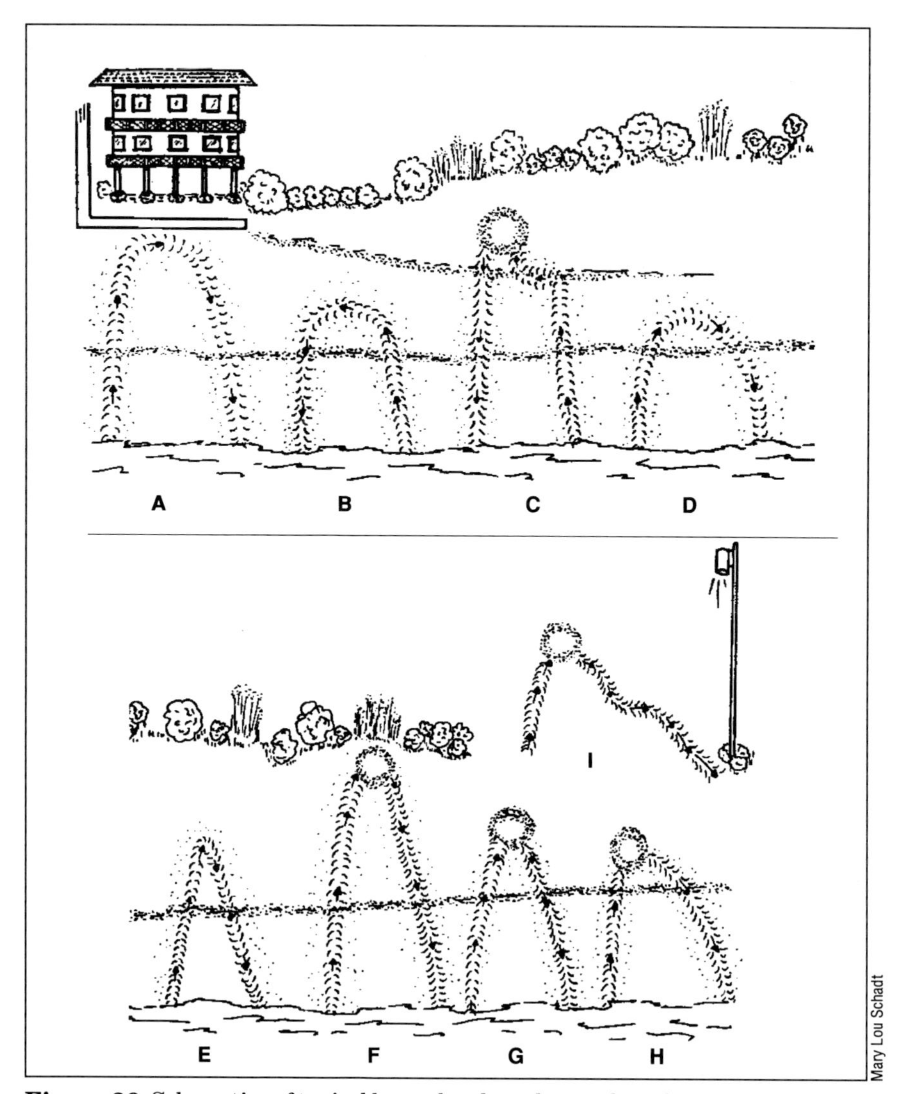

Figure 36. Schematics of typical loggerhead turtle crawl tracks.

horizon lights of mainland development. This type of crawl is very common on most of Sanibel Island because of the island's configuration. Mainland lights, from Fort Myers Beach south to Naples, are situated to the left, or, on the Island's eastern end, almost ahead of a postnesting female returning to the water. Sometimes, a flashlight or Coleman lantern being carried by a beach user will have the same attractive effect. Crawl I is the result of nearby artificial illumination, perhaps a bright mercury vapor security light. Following the nest concealment stage, the turtle turned toward the water, but her sea finding cues were overpowered and she headed off in an angled crawl in the direction of the bright light. Turtle I reached the surf, but only after laboriously moving an unnecessary distance. I have observed these erratic crawls over the years, some resulting in a return crawl hundreds of feet in length. Once, a loggerhead nested on the western end of Sanibel Island in front of a small six unit resort. Following normal turning from the completed nest site and starting towards the water, this turtle looped inland, attracted by the glare of street lights several hundred feet away on the road fronting the building. Before this struggling individual safely reached the sea, she had completely encircled the structure. It is possible to take a flashlight, stand in front of the animal, shine the light on the beach surface ahead of her, and literally lead a postnesting female from the nest site in any direction desired.

Another very unusual loggerhead emergence comes to mind. One summer, back in the late sixties, a young honeymooning couple from Lakeland, Florida, decided to camp in a tent on the Gulf beach near Redfish Pass, on Upper Captiva. Late at night their romantic solitude ended and their nerves were shattered, when a large loggerhead turtle made sudden contact with their tent, knocking it down on them as she moved up the beach!

10
Nest Site Tenacity

Identification of individual turtles through tagging programs, has helped us to clearly understand several important aspects of sea turtle biology, including the group's remarkable nesting site fidelity. That these reptiles can return with such accuracy after a two or three year absence is uncanny and an intriguing part of sea turtle biology.

The Caribbean green turtles, many of which nest at Tortuguero, Costa Rica, are one of the most studied sea turtle populations in the world. Tagging studies which were initiated there by Archie Carr have been conducted on this beach for over three decades and thousands of these turtles have been tagged. There is no record, at least none published, of even one of these tagged green turtles nesting anywhere else—the epitome of sea turtle nest site tenacity.

When tags were first being applied to Sanibel Island *Caretta*, I assumed that loggerheads tagged on Sanibel would nest only on Sanibel since Sanibel-tagged loggerheads were returning for renesting with regularity. Each year about 40 percent of the adult turtles encountered on the beach had been tagged on the island during earlier years. But in 1972, one particular animal provided an exciting deviation from my assumption.

On May 28, 1968, I applied one of Archie Carr's tags (C 429) to a loggerhead immediately after she finished nesting on Sanibel Island, nine miles west of the Sanibel Lighthouse. This animal was not observed again until it was discovered nesting on June 25, 1972, immediately south of Melbourne Beach, Brevard County, Florida. This event which I later published in the scientific journal, *Herpetologica*, was the first documented case of a sea turtle displaying such an unusual nesting site relocation. The same turtle nested on both sides of the Florida peninsula! In the years since this occurred, a loggerhead which had been tagged by Frank Lund on Jupiter Island, Martin County, during a false crawl, was discovered nesting on Sanibel three years later. Evidence has mounted that some loggerheads, which periodically nest in the study area, are not as site specific as was presumed.

In the early seventies, when the summer field work was expanded to

Patrick D. Hagan

Plate 18. This loggerhead turtle was tagged on Jupiter Island, on Florida's east coast, in 1973. On July 20, 1976, Paul Zajicek and Pat Hagan encountered the turtle nesting on Sanibel Island.

include virtually all nesting beaches in the subregion, larger sampling of the nesting population provided a much clearer picture of what was really going on. Most individuals, such as loggerhead CR 140, remained completely faithful to the Sanibel beach, nest after nest. I now know that some females may drift along the coastline, coming ashore at different and distant islands for egg deposition. This movement may be casual in nature, or it may be intentional and consistent with the life-style of the individual. Some of the islands are extremely close together, only separated by narrow passes a few hundred feet in width. This continuity of the shoreline in the subregion must be considered when one talks about inconsistencies in site tenacity.

One individual tagged on June 4, 1973, (CR 439) at my Cape Romano study beach, was discovered renesting forty-two nights later (July 16) on Sanibel Island, 9.9 miles west of the Lighthouse. This was a point to point, straightline move of some fifty-two miles. More recently a loggerhead tagged in the early morning of June 15, 1986, (CR 5142, SAN 132) 2 miles west of the Sanibel Lighthouse, was discovered on July 11, 1986, nesting on the north end of Casey Key, Sarasota County, about fifty-six miles north of the initial deposition and tagging location. Many other individual loggerheads have made similar shifts of nesting beaches in both internesting and interseasonal nesting landings. Why the majority of marked turtles will tenaciously return to the same beach, and others will not, is another interesting, but little understood segment of this species' life history.

One factor which may have implications to nest site shifting involves some form of disturbance which negatively impacts individuals, such as interaction with humans who utilize the same beaches at night, when strolling, shelling, and surf fishing, etc. Other adverse conditions result in immediate termination of the nesting effort and, therefore, a false crawl. Once back in the water, the individual may move parallel to the beach, crossing passes to make another nesting attempt on some other island; however, this does not explain the frequent wide distances between the nesting emergences of some females.

Handling turtles for weighing, measuring and tagging, especially when an individual must be turned upside down for restraint, has to be a traumatic experience for the turtle. I have turned turtles only when that course of action was necessary for some specific purpose. One would imagine that such treatment would discourage a loggerhead and send her off to some less troubling area, although evidence on Sanibel Island suggests that this has not been the case. Loggerhead turtle CR 140, discussed in the chapter concerning fecundity, is a classic example of an individual which was somewhat roughly treated yet continued to maintain remarkable integrity in the fidelity of her Sanibel Island nest sites. I recall another individual, CR 126, in 1971 who was turned, weighed, measured and tagged during a false crawl. This turtle was discovered and physically examined to confirm its identification three more times the same night. Her second and third emergences also were non-nesting, but her fourth try resulted in a successful nest. This individual was encountered nesting twice more that particular summer and has been recorded renesting on Sanibel over the years. More recently, on June 24, 1989, a loggerhead aborted her nearly completed egg cavity which was situated in front of a resort building. The animal was half-way to the water, apparently having been disturbed just moments before by children playing a few yards ahead of her on an illuminated sun deck. Hard, wet sand made our restraining device useless, so to prevent her escape, the loggerhead was quickly turned upside down. A flipper tag and a carapace tag (CL 144, SAN 175) were applied and shell measurements were taken. Two and one half hours later, as we made our return pass along the Sanibel shoreline, this same turtle crossed the beach in front of us and deposited eighty-four eggs one hundred yards to the west of her initial emergence.

The majority of loggerheads nesting on the coast of the subregion exhibit a precision in their nest site fidelity. There are some independent individuals, however, who apparently use subregional beaches for purposes of egg deposition in what seems to be a hit-or-miss, rather carefree, style. Such data indicate to me that *Caretta* is not unswervingly faithful or dependable in terms of nesting behavior or nesting beach site specificity.

11 Fecundity

Egg counts are made when nest monitoring procedures include the need for such data and whenever a clutch is relocated close to its original location or transplanted some distance away into a hatchery. The total number of eggs contained in loggerhead clutches, which I will use are from accurate counts and not from counts that were made as eggs fell from the cloaca of nesting females. Such deposition counts which are done visually have proven to be inaccurate.

The technique I have customarily used to determine clutch size is quite simple. Today, this method is used only for nest contents to be relocated that were deposited moments before they are counted. One person excavates the nest chamber by hand and carefully removes the eggs, while another individual positions the clutch nearby in groups of ten. When all eggs have been removed from the nest and so arranged, they are counted and this total is recorded. When they are replaced into the chamber, they are counted again, but by two's, for verification. Some researchers use hand tally counters and record eggs as they drop from the ovipositor into their hands which are held above the cavity, underneath the turtle. They report good results from this technique.

Marine turtles are known for their amazing ability to produce large egg complements. This element, coupled with multiple nesting behavior, makes sea turtles the most prolific of the Reptilia. Through time, evolutionary processes have developed to enhance the survivability of species in several key ways. In the case of marine turtles, one aspect of this enhancement is the production of profuse numbers of gametes which contributes to their long-term survival. Parental care is not an attribute common to reptiles, so to overcome natural attrition these and many other species produce prolific numbers of progeny to ensure survival of the species.

During most years of the Southwest Florida study, loggerhead turtle clutch sizes were counted routinely. Documentation of clutch data is contained on field data sheets which are completed for each individual turtle or, if specific identification of a nesting female is unknown, for each nest that has been transplanted or examined for some reason.

Clutch size ranges which I provide include data from all nests—both parentage known and unknown lineage complements. On rare occasions, nesting females were frightened and aborted in-progress deposition. Clutch sizes are not included for individual turtles where such unusual events occurred, and only numbers from normally produced clutches are included.

Typically, when a nesting loggerhead is found on Sanibel Island, it is observed from a distance until the supervising field worker determines what level of nesting behavior is involved. If covering activity is in progress, marks are usually made on the beach surface, or small sticks are stuck nearby in the sand, to allow easier later location of the eggs by triangulation from the marks or sticks.

While many individual loggerheads were found one or more times in a season during their multiple nesting landings, one individual loggerhead was outstanding in her nest site tenacity and egg productivity. Turtle CR 140 was originally tagged on Sanibel Island on July 27, 1970, approximately one mile west of the Sanibel Lighthouse. On this visit, she deposited a clutch consisting of 159 eggs. CR 140 was observed again, contrary to the expected rule for the cyclic nesting for Southwest Florida loggerheads, the next summer on July 13, 1971, 2.3 miles west of the Lighthouse. On this, her only observed reproductive visit for the year, the nest contained 149 eggs.

In 1973, CR 140 established a documented level of egg production that, I'm sure, represents a Gulf of Mexico, if not a western Atlantic, record for the species. Seldom has any sea turtle worker verified total egg productivity of an individual loggerhead, as a result of multiple nestings, much beyond three or four nests per season. CR 140 provided the opportunity for documentation of a large female loggerhead's egg production on a scale that, to my knowledge, has never before appeared in the literature relative to the reproductive behavior of *Caretta caretta.* At 0025 hours on May 24, 1973, CR 140 was discovered on the Sanibel beach 2.3 miles west of the Lighthouse. This first clutch contained 174 eggs which were transplanted into a natural beach hatchery compound located on the Lighthouse property. At 2320 hours on June 4, CR 140 was encountered 4.7 miles west of the Lighthouse where she deposited 162 eggs, which were also transported to the hatchery. Later, on June 18, at 0050 hours, CR 140 was found 3.1 miles west of the Lighthouse and the clutch contained 152 eggs which were transferred to the hatchery. On July 5, at 0150 hours, CR 140 was discovered nesting 2.3 miles west of the Lighthouse. The 145 eggs from this clutch were relocated into the protected hatchery.

Following the fourth nesting, and since the time period was just about at mid-nesting season, I increased vehicular patrol activity. Two vehicles, with all-night operators and volunteer forces, were put in the field nine nights after the fourth nesting of CR 140. At 2300 hours, on July 15, CR 140 was again on the Sanibel beach, 3.8 miles west of the Lighthouse, nesting for a fifth time. This complement contained 144 eggs (two were atypically small in size) which were relocated into the hatchery.

Nine nights later patrol coverage activity was again increased with two field groups operational, and CR 140 was discovered on the beach a sixth time 1 mile west of the Lighthouse on July 29, at 0015 hours. On this unprecedented sixth nesting, this loggerhead deposited a clutch totalling 143 eggs. These were also relocated into the hatchery compound at the Lighthouse. Nine nights later I launched another saturation patrol effort, but CR 140, or "Myrtle" as she is fondly remembered, has never been observed again since that remarkable sixth nesting visit in the summer of 1973.

With her six egg deposition visits to Sanibel Island, CR 140 provided a considerable amount of new life history information. Her documented production totalled 920 eggs, she demonstrated outstanding nest site tenacity during the summer, and provided interesting data on egg and hatchling sizes from the various multiple clutches. These data are discussed on page 37. Many other loggerheads of the subregion have been observed nesting four, and even five, times in one summer, but none ever closely approached CR 140 in total eggs produced for one nesting season.

CR 140 was a large loggerhead turtle. This specimen was measured and weighed only once in 1973, on May 24. Straightline measurements were: carapace length, 41 inches (104.2 cm); carapace width, 30 inches (76.2 cm) and cranial width 8 inches (20.3 cm). This animal weighed 325 pounds. Loggerhead turtle eggs from Sanibel Island average 41.0 grams in weight. When the individual egg weight is multiplied by the 920 normal eggs she deposited, CR 140 produced about 83 pounds of eggs in 1973—or more than 25 percent of her total body weight.

It is noteworthy to review the locations on Sanibel Island where CR 140 landed during the 1971 and 1973 nesting seasons. Her only known nesting in 1971 was observed 2.3 miles west of the Sanibel Lighthouse and, almost two years later, in 1973, she was encountered within a few hundred feet of that earlier site. On her fourth nesting she was found within the same statistical zone, a hundred yards away from her first nest of the season. Throughout the 1973 nesting season, CR 140 "homed" to the eastern end of Sanibel Island and concentrated her six nesting sites to a 3.7 mile section of beach—a remarkable example of nest site fidelity in *Caretta!*

It should be noted that CR 140 produced egg groups that diminished in number with each subsequent nest, unlike the Cape Romano turtle, CR 441, where clutch sizes increased as her multiple nests were deposited. This variation in the overall population is commonplace and may be individually

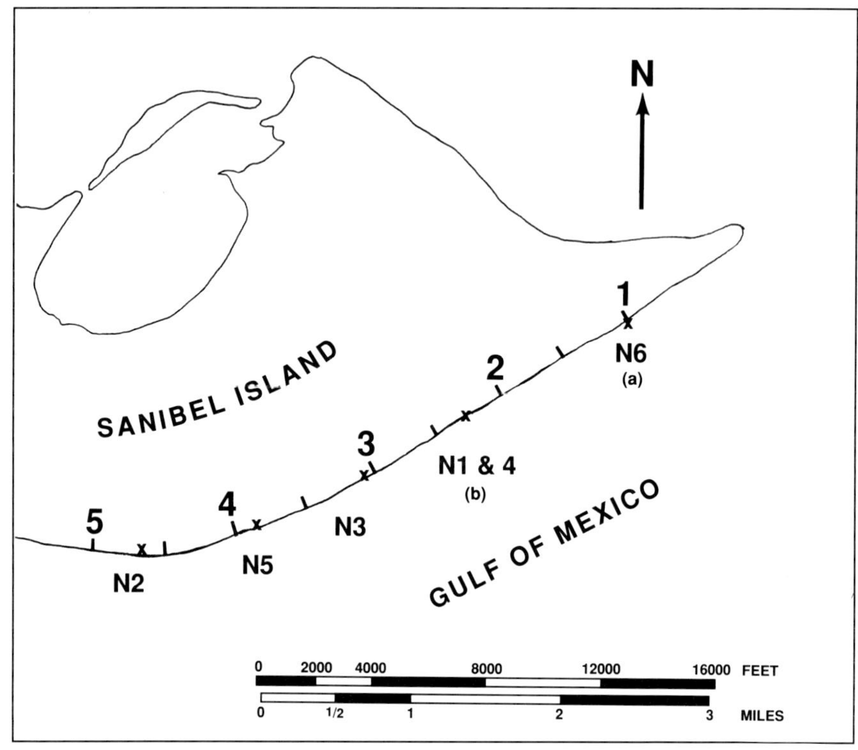

Figure 37. Nesting sites of CR 140 on eastern Sanibel Island in 1970, 1971 and 1973. Numbered sections of the beach (1-5) identify the one-half mile long statistical zones. N 1-6 (x) mark the sequential nests of 1973. Sites N 1 and 4 are within the same one-half mile section of beach as was the nest of 1971, (b). This loggerhead's observed nesting in 1971 (a) was also located within the same four mile long section of beach.

determined. Most of the data I have collected on egg production over the years indicate that, in general, there is a decreasing trend in the clutch size as each subsequent multiple nest is produced.

In the summer of 1969, loggerhead turtle CR 111 was encountered on the Sanibel beach four times depositing a clutch of eggs each time. The animal produced progressively fewer eggs each time she nested until the fourth clutch. A two day interval between clutches three and four was recorded—a remarkably irregular schedule!

During development, eggs begin to calcify about seventy-two hours

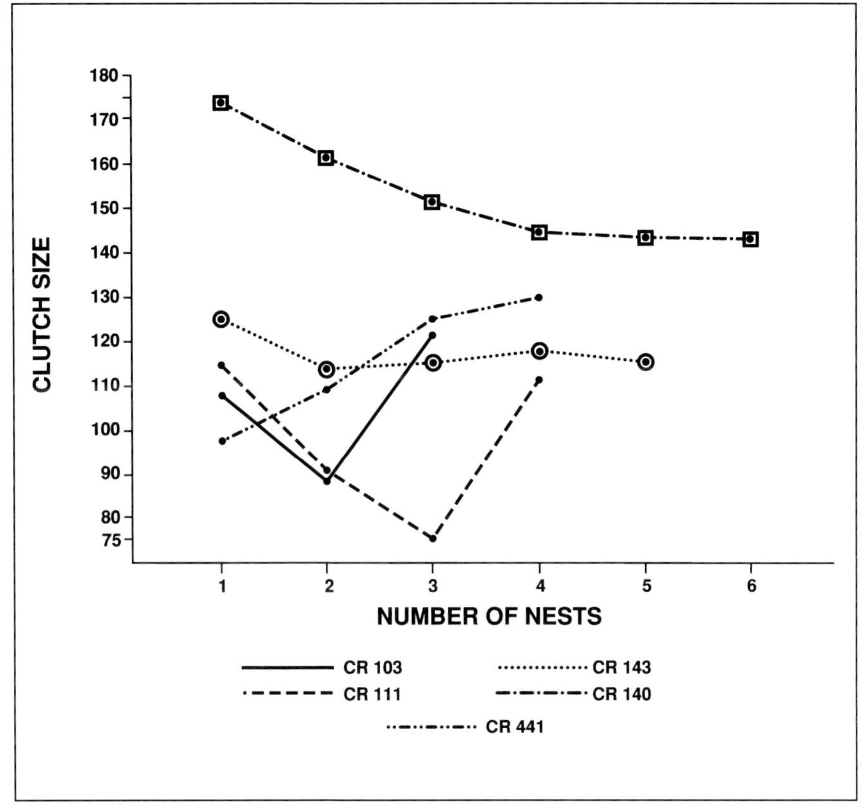

Figure 38. Total egg production of selected loggerhead turtles making three or more observed in-season nestings on subregional beaches. 1964-1989.

after ovulation and the minimal calcification of *Caretta* eggs is usually complete ten days following ovulation. In female sea turtles, live spermatozoa is retained in a viable state internally for some time after copulation and are known to occupy glands near the shell forming region of the oviduct. Eggs are fertilized for the next oviposition event just prior to the start of shell buildup.

It would appear that CR 111 somehow divided clutches when the remarkable fourth clutch followed the third only two nights later. Most of the eggs of the fourth nesting would have been premature, or may not have been fertilized, based on the above outlined development schedule. The viability of each clutch also suggests that clutch four may have contained a

Clutch	Date deposited	Eggs produced	Live hatchlings produced	Percent hatched
1	5/26	116	104	90
2	6/8	89	84	94
3	7/11	76*	64	84
4	7/13	111	31	28
Totals		**392**	**283**	

*This clutch also contained two, small yolkless eggs that are not included in the total.

Table 2. Clutch sizes and hatching success of eggs deposited by CR 111 in 1969. Based on the ordinary internesting interval for subregional *Caretta*, this individual probably made an unobserved nesting visit between clutches two and three.

number of mature fertilized eggs which should have been deposited with egg group three, but for some unknown reason were not. All of the eggs were moved and transplanted to a hatchery at the Sanibel Lighthouse within three hours of deposition. The sizes and success of each clutch are shown in Table 2.

12
Egg Size and Shell Thickness

Although they appear to be, loggerhead turtle eggs are seldom perfectly round. Measurements of their eggs were routinely recorded during the 1970's on Sanibel Island in conjunction with major egg protection efforts. Typically, a series of eggs from the top and bottom of each clutch excavated for transplantation, usually five from each elevation, were measured at the time they were reburied into a hatchery enclosure. The bottom eggs represent the first eggs deposited while the top ones were among the last produced. The egg series were measured with calipers, and two diameters were measured at right angles to each other. There was no depression of the egg surface and accurate diameters were recorded. The eggs produced by loggerheads on Sanibel Island in 1976, based on a combined (both top and bottom eggs) sample size of 300 eggs averaged 41.1 by 42.2 millimeters.

When eggs are deposited, they are somewhat flaccid; the shells' surfaces depress as they contact one another in the nest cavity. When held, finger pressure on the shell will create a depression or dimple where the greatest pressure is exerted. These hollows remain for three or four days until water uptake by the egg creates a rise in interior pressures which force the dimple out in fertile eggs as the eggshell tautens. Infertile eggs remain flaccid.

The eggs of subregional *Caretta*, more specifically those of the Sanibel Island nesting population, have been found in a wide variety of configurations and sizes. Some females have produced, along with normal eggs, masses of material that resembled folded sheets of parchment. Others have produced elongated multi-yolked eggs up to 37.4 by 125.0 mm, and some individuals have deposited eggs that were fastened together by thin string-like connections. Over the years, I have examined many clutches which contained diminutive yolkless eggs. Some sea turtle researchers suggest that these small eggs and, perhaps some of the atypical sizes and shapes, are indicative of clutch components, or surplus egg material, which are being purged from an individual turtle's reproductive system at the termination of her respective nesting season. This may well be the case for I have often

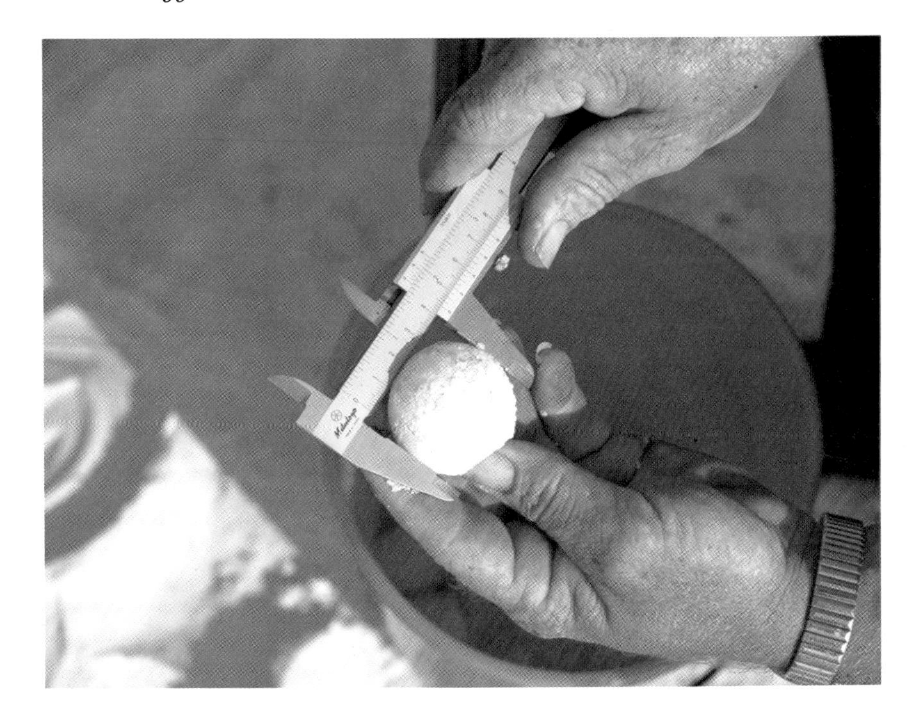

Plate 19. Eggs are measured with vernier calipers. The left thumb has depressed the flaccid egg to tighten the surface between the parallels of the calipers and produce an accurate measurement of the egg's diameter.

discovered small yolkless eggs and shell debris in nests after a known tagged turtle had deposited her third to fifth nest. The fifth clutch of CR 140, which contained two small eggs, is contrary to this assumption, for she returned and deposited a sixth complement which contained only normal eggs.

In reviewing the Sanibel Island egg size data, I find that the smallest loggerhead turtle egg I have ever inspected and measured was 8.0 by 10.0 mm in diameter. In contrast, the largest near-spherical egg I have observed in a Sanibel Island *Caretta* nest, and also have retained in my personal collection, measures 53.0 by 60.0 mm.

In 1976, to better comprehend the loggerhead turtle egg and document other aspects of its composition, I initiated a short-term, one nesting season, study of egg shell thicknesses of eggs produced on Sanibel Island. A comparative series of eggs was also collected from the Florida Atlantic coast, in Brevard County, to the north of Sebastian Inlet State Recreation Area. Fresh eggs were collected from a series of nests during, or immediately following, deposition. Eggs were selected from nests that were

Plate 20. A most unusual photograph . . . connected eggs stopped at the instant of oviposition. The normal mucous that flows along with each egg is visible, streaming from the upper egg.

threatened because of their close proximity to the water's edge and where, because of site conditions, it was reasonable to assume that the clutch would not have survived to full term development. The remaining eggs of these selected clutches were relocated higher on the beach to improve odds for their survival. Ten eggs from ten Sanibel loggerhead nests represented the sample size utilized in the data base. A series of ten similarly jeopardized clutches from the Brevard County region provided a sample size of 100 eggs too. The collections from both locations represented five eggs from both the top and bottom of each nest. Again, the samples included eggs that were deposited first (bottom) and eggs that fell last (top).

The procedure used in preparing eggs which were included in this study was as follows:

1. All eggs were hand washed under faucet water pressure until all adhered sand particles or other matter was removed.
2. An Ohaus triple beam balance was used for weight determination

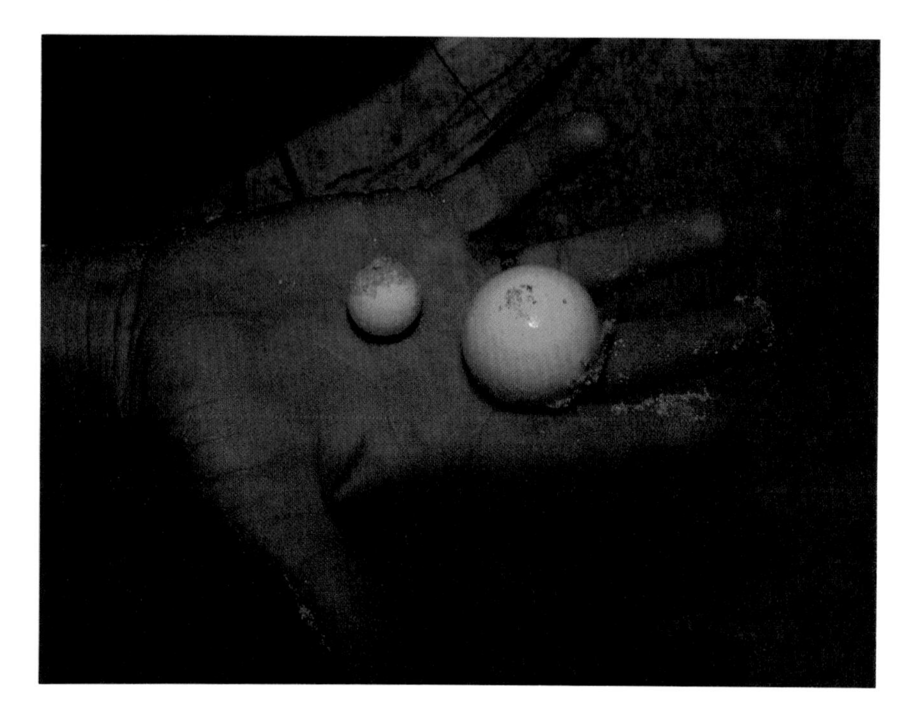

Plate 21. A normal egg on the right and a diminutive, yolkless, egg to the left.

after each egg was washed and air dried until no moisture was visible on the outer egg shell. Weights were recorded in grams.

3. Vernier calipers were used to measure two diameters of each egg at right angles. Finger pressure on each egg altered location of the normal egg surface dimple and, therefore, this feature did not in any way influence accurate diameter determination. Egg surfaces were not depressed between the parallels of the calipers during measuring. Measurements were given in millimeters. Mean dimensions were determined by averaging the minimum and maximum of each egg.

4. Each egg was opened, and its contents emptied. The shell was gently washed under faucet water pressure. A micrometer was used to measure shell wall thickness at three random locations. Egg shell thicknesses were reflected in thousandths of an inch and were averaged from the three measurements.

These data are summarized in Tables 3 and 4 and were analyzed using

Turtle	Date Eggs Collected	Total Clutch Size	Straight Carapace Length (in inches)	Weight (in pounds)	NEST TOP Mean Egg Diameter	NEST TOP Mean Egg Weight (in grams)	NEST TOP Mean Shell Thickness	NEST BOTTOM Mean Egg Diameter	NEST BOTTOM Mean Egg Weight (in grams)	NEST BOTTOM Mean Shell Thickness
CR 610	6 / 17 / 76	119	37.0	245	41.9 × 43.6	41.8	.265	42.9 × 43.7	44.0	.276
CR 1011	6 / 21 / 76	115	36.5	230	43.5 × 44.7	47.8	.275	43.4 × 44.4	46.2	.270
CR 1780	6 / 21 / 76	107	36.0	235	40.5 × 42.1	38.0	.268	38.8 × 45.1	39.0	.280
CR 1781	6 / 26 / 76	146	40.0	285	43.4 × 44.3	46.1	.253	43.3 × 44.8	46.0	.244
CR 258	7 / 04 / 76	82	35.5	210	41.2 × 42.2	39.8	.280	40.9 × 41.7	36.6	.247
CR 1783	7 / 04 / 76	127	33.5	195	39.1 × 39.7*	33.0*	.289	39.8 × 41.0	35.7	.219
CR 391	7 / 11 / 76	111	38.0	295	43.8 × 44.3	45.5	.308	43.3 × 44.0	45.2	.290
CR 601	7 / 11 / 76	75	35.0	200	38.0 × 38.3	30.4	.251	38.7 × 38.9	30.3	.236
CR 1782	7 / 12 / 76	110	32.25	225	39.2 × 39.6	33.4	.244	40.0 × 40.2	35.4	.244
CR 1012	7 / 13 / 76	127	38.5	270	41.1 × 41.5	39.1	.243	41.1 × 41.7	39.6	.240
Averages		**111.9**	**36.5**	**239**	**41.1 × 42.0**	**39.4**	**.267**	**41.2 × 42.5**	39.8	**.254**

*One yolkless egg included in sample. Size—33.4 × 34.0, weight 20.0, thickness .296.

Table 3. Clutch size, egg dimensions, egg weights, and egg shell thicknesses from ten *Caretta* egg clutches from Sanibel Island, Lee County, Florida. 1976.

Turtle	Date Eggs Collected	Total Clutch Size	Straight Carapace Length (in inches)	NEST TOP Mean Egg Diameter	NEST TOP Mean Egg Weight (in grams)	NEST TOP Mean Shell Thickness	NEST BOTTOM Mean Egg Diameter	NEST BOTTOM Mean Egg Weight (in grams)	NEST BOTTOM Mean Shell Thickness
SI 271	6 / 27 / 76	134	33.75	41.3 × 42.6	40.1	.268	41.9 × 43.5	42.2	.275
CR 1751	6 / 28 / 76	124	33.5	43.6 × 44.0	46.0	.317	43.4 × 43.9	46.1	.251
CR 1756	6 / 28 / 76	148	34.5	41.4 × 42.6	41.3	.203	42.6 × 43.7	44.4	.267
SI 210	7 / 08 / 76	105	38.0	41.8 × 42.3	40.5	.265	43.4 × 44.3	45.8	.233
CR 1348	7 / 10 / 76	134	35.75	39.0 × 40.6	32.3	.262	41.7 × 43.5	39.4	.252
CR 1769	7 / 10 / 76	101	40.0	42.7 × 43.5	44.4	.277	42.5 × 43.8	44.1	.239
CR 1382	7 / 11 / 76	133	40.0	41.2 × 42.7	39.9	.257	42.1 × 43.3	42.3	.262
CR 1385	7 / 12 / 76	123	37.0	44.6 × 45.2	49.5	.280	44.9 × 45.7	51.0	.263
CR 1386	7 / 12 / 76	100	34.5	41.0 × 41.7	38.8	.222	40.4 × 41.9	37.8	.230
CR 1387	7 / 12 / 76	107	35.0	43.2 × 43.8	44.8	.278	43.3 × 44.2	46.3	.251
Averages		**120.9**	**36.2**	**41.9 × 42.9**	**41.7**	**.262**	**42.6 × 43.7**	43.9	**.252**

Table 4. Clutch size, egg dimensions, egg weights, and egg shell thicknesses from ten *Caretta* egg clutches from Brevard County, Florida. Adult weights were not recorded. 1976.

the Pearson-Moment coefficient and the Spearman-rank Correlation coefficient. Only the following relationships were statistically significant.

Sanibel Island:

1. There was no relationship between the bottom egg weight and the size of the clutch.
2. There was a relationship between the length of the carapace and the weight of the turtle. In other words, longer turtles weigh more.

3. There is a correlation between turtle weight and the following: top egg width, top egg length, top egg weight, bottom egg width, bottom egg weight. In other words, the larger the turtle the larger the egg.

Even though there was not a significant relationship between clutch size and the thickness of the eggs, it was interesting to note that both coefficients were negative which would indicate that as the clutch size increased the egg thickness, both top and bottom, decreased. It was also interesting to note that these were the only negative values related with clutch size.

Because there was no common factor to relate the Sanibel Island and Brevard County samples, as there might have been had the same turtle laid eggs in both places, a correlational analysis was not performed for this requires bivariate data. Consequently, a t-test was performed to determine if there was a difference between the means of the two groups.

There was a statistical difference between the top shell thicknesses and the bottom shell thicknesses on Sanibel Island. The p-value for this test was an unimpressive .1223. In other words, there was a difference between the means, but there is a 12.23% chance of being in error by saying that there is a difference.

Brevard County:

1. There was a correlation between clutch size and bottom shell thickness. It was interesting to note that although there were no other correlations with clutch size, that the following indicated a negative trend: carapace, top egg width, top egg length, top egg weight, top shell thickness, and bottom egg weight. This means that as the size of the clutch increased the other variables decreased in size. The top egg length correlated with the top egg weight, bottom egg width, bottom egg length, but with neither egg shell thickness.

The only really significant statistical difference was between the shell thickness of the bottom eggs from Brevard County and the top eggs from Sanibel Island. This indicates that there is a difference between the thickness of the shell; however, a great deal of faith cannot be placed on the results because of the small sampling. There is a good chance of sampling error existing; thus, more data and more tests should really be done before any sound conclusions can be reached.

13 Clutch Sizes

The total number of eggs deposited in each individual nest cavity, or clutch size, produced by subregional loggerheads during each of their in-season nesting landings were usually tallied during the early years of my monitoring program. More recently, eggs from clutches which are being relocated, for one reason or another, are counted and appropriately recorded. The contents of nests from both known (tagged) individuals and counts from nests of unknown parental lineage, which were within the limits of the study area, were counted. Counts were made as part of the nest protection program, or were included in some other facet of the generalized monitoring aspects of summer operations. Only accurate, double checked as previously discussed, counts of egg groups are included in the following data. The largest sample size, because of the long-term uninterrupted effort, is available from Sanibel Island. In my earlier published work, *The Marine Turtles of Sanibel-Captiva Islands, Florida* (1969), I presented the average clutch size for Sanibel Island *Caretta* as 110 eggs. Some researchers working with this species on the eastern seaboard of the United States have determined that the mean loggerhead turtle clutch size is in the neighborhood of 120 eggs, with specific averages depending on what nesting population is under consideration.

The greatest range in egg clutch sizes for the Southwest Florida loggerhead turtle nesting population was recorded on Sanibel Island, with clutch sizes ranging between 27 (1976) and 181 (1978) eggs. Clutch sizes are variable from nesting season to nesting season, and an examination of clutch sizes from several different years clearly illustrates the variability. For example, in 1971 clutch sizes ranged from 91 to 140 eggs and averaged 114 eggs. Two years later, in 1973, clutch sizes ranged from 85 to 174 with an average complement containing 120 eggs. Two years thereafter, in 1975, the average clutch size was much lower at 106, and ranged between 48 and 146 eggs. Examining twenty years of gross data, the average clutch size for this population is 108 eggs per nesting. This is slightly lower than what I had assumed to be an average clutch size from records which had been based on data collected in the early years of my field studies. The years

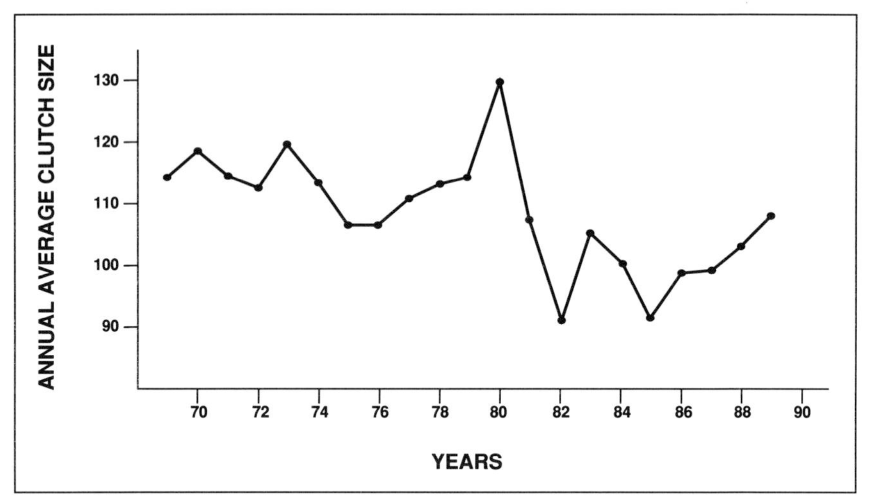

Figure 39. Annual average clutch sizes produced by Sanibel Island loggerhead turtles. 1969-1989.

1969-1989 were selected because during those seasons clutch size determination was continuous. These average clutches are shown in Figure 39.

Smaller clutches than the minimum recorded clutch given previously (27) have been observed and recorded by my associates or me on Sanibel Island; however, in my opinion these were not the result of natural, undisturbed, egg deposition processes. Such egg groups have been purged from the above statistics.

The specimen that produced the minimal clutch of 27 eggs, turtle CR 1783, was a small individual with a straight carapace length of 33.5 inches (85.09 cm), who weighed 195 pounds. The maximum clutch, a 181 egg complement, was deposited by turtle CR 1466, a large loggerhead with a straight carapace length of 40.0 inches (101.6 cm). This specimen was not weighed because I had discontinued weighing adult loggerheads by 1978.

14
Incubation Period

For purposes of incubation reptilian eggs are buried in holes in the ground, placed inside some form of vegetative debris, or underneath deteriorating materials which contact the ground, by the egg-laden female. Embryonic development is then regulated by ambient air and soil temperatures as well as moisture levels in the surrounding incubation medium. Developmental timetables, from egg deposition until pipping of the eggshell by a full-term hatchling, are dependent on the effects of temperature and moisture for embryonic growth. This period of incubation, unlike the more specifically timed mammalian gestation, can be shortened or greatly extended for eggs produced by individuals of the same species. The developmental term of egg clutches in subregional *Caretta*, even groups produced in the same season by the same female, can vary widely because of these controlling environmental factors. I use the term, incubation period, specifically to identify the elapsed time required for a hatchling, or an entire clutch, to develop. This does not include the additional time required for a group of hatchlings to undermine their developmental chamber and finally reach the surface of the beach. The slow movement upward through the soil can add anywhere from two to six days to the time required for a united group of young turtles to reach the beach surface and freedom as they escape their dark developmental chamber.

Incubation of subregional *Caretta* eggs, those in natural clutches that remain in situ and undisturbed, have ranged from forty-nine to seventy-two days. Loggerhead eggs which were removed from natural egg cavities, and transplanted into a protective hatchery compound, had a similar range in the duration of their development. Sample clutches which were monitored to document incubation period time spans were checked by both sight and sound.

The average incubation period over the decades has been fifty-six days for the egg complements of subregional *Caretta*. In the 1970's, the individually marked nests, which were being monitored on Sanibel, were examined nightly beginning forty-five days post-deposition. The long metallic probe of an automotive mechanic's stethoscope was inserted several inches into the soil above and to the side of the egg chamber being examined. With the sensitive device, an examiner could clearly hear the persistent scratching and squirming of any hatchlings that had broken free of their eggshell. Later, evidence of soil cracks, surface depressions forming immediately above the location of the eggs, or tiny turtle tracks in the sand would indicate that hatchlings had vacated the chamber. A few times over the years, we were fortunate enough to be on hand to witness the exciting exodus of little loggerheads from their nest.

On rare occasions a high percentage of the eggs in a clutch will be infertile. I have opened a few natural undisturbed nests which were situated in ideal beach conditions, and discovered that all eggs were addled and apparently had never started to develop. It is just about as rare to find a 100 percent success rate, where all eggs hatched and all hatchlings exited the nest. A typical nest usually has an infertile rate ranging from 3 to 20 percent.

On Sanibel Island, typical egg clutches which are monitored and not manipulated in any way prior to hatchling emergence, produce at an 86 percent success level. The beach hatcheries which were once operated on Sanibel resulted in a success average of 82.6 percent.

15
Embryonic Development

Free of their mother's body, yet confined to a similar dark crowded space in the earth, the recently deposited eggs continue the amazing transformation of embryological development. Subtle changes have already begun inside the fertile eggs by the time they are nestled in the dark egg cavity.

Embryonic development in loggerhead turtles has been divided into thirty-one stages, with the first six occurring pre-ovipositional. Prior to deposition, when they are mature eggs but still in the lower end of the oviductal passageways, the interior development of an egg has attained a blastula state—only a single layer of cells is present. The level of development within a loggerhead turtle egg at the moment of deposition is late gastrula, or stage six. The embryo is formed into a cup-like body of two layers of cells that enclose a central cavity.

Once an egg becomes immobile in its incubation medium, the vitelline sac, or embryonic disk, containing the gastrula and nearby membrane, begins to rise to the upper part of the sphere where it adheres to the membrane which lines the inner shell wall. The vitelline membrane separates the yolk and embryo from the albumen. When adhesion to the shell membrane has occurred, a chalk-white spot develops on the top of the egg at the location where the membrane has adhered. This white area continues to enlarge and does not terminate until the vitelline membrane has completely encased the inner surface of the egg shell. By the time complete adhesion has occurred, the egg has become totally chalky white. Should the embryo succumb at the outset, the vitelline sac will shrink. The egg then becomes very flaccid and discolored, acquiring the general appearance of being unfertilized. If we were able to view the profile of a fertile loggerhead turtle egg a few days following oviposition, we would observe a changing sphere with an obvious division of color diametrically around its outer wall. As mentioned, this change begins at the top of the egg and progressively moves down radially until the egg color has become completely altered. Barring death of the embryo, the complete color metamorphosis takes about ten days.

When actual egg deposition was not observed, we can calculate a relatively accurate estimation of the clutch's age by visual observation of the degree, or percentage, of this very obvious color delineation. The change in pigmentation is quickly determined by looking at an egg and judging the extent of the color division. For example, a spot of chalk-white on the top of the egg equals twenty-four hours since deposition; whereas, an egg with a profile having one half of each color would be determined to be six days old. Estimation of age between six and ten days relates to the percentages of new color in the lower hemisphere of the egg.

Early on, the vitelline sac is important for balanced gas exchange through the permeable egg wall. Oxygen is supplied via the sac from outside the egg to the embryo and carbon dioxide is similarly transmitted externally, passing through the egg shell. The rates of oxygen intake and output of carbon dioxide is believed to be minimal during establishment of the vitelline system. Later, in about the last twenty days of development, the consumption rate of oxygen multiplies, the vitelline life support system is no longer adequate, and allantoic circulation becomes the source of the oxygen.

The allantois is the major respiratory and excretory membrane in the advanced developmental stages of the embryo. As this membrane enlarges and becomes more complex in its oxygenated blood supply and venous circulation, it assumes full responsibility for proper gas exchange. The membrane also functions as a waste storage system in the final trimester of development.

Under certain circumstances, the vitelline membrane adhesion may be a critical factor which can reduce the viability of eggs should a clutch have to be moved. I am of the opinion that once the vitelline membrane connection is made to the egg wall, it is imperative that the egg not be turned incorrectly, for this can result in tearing of the vitelline tissue from the shell and cease development. Eggs which I have transplanted any distance were moved and reburied within a maximum time frame of twelve hours. During the years when I mass transplanted eggs into beach hatcheries, they were always in the ground within six hours of deposition. Now, the occasional nest that in my judgment is in danger, regardless of its age, is relocated to a safer site near the original nest cavity with extreme care. Eggs can be spun on their axis without any apparent damage, but when they are rotated the other way, rolled upside down beyond the twelve hour threshold, the embryo is killed. Some sea turtle conservationists have found that eggs which are close to the end of the incubation period can be safely moved without a real threat to their viability. Several years ago, in an attempt to restore loggerhead turtles on some of their historical nesting beaches, near full-term eggs were collected from South Carolina and transported to Virginia. There, egg chambers were excavated on the selected Virginia beach and the South Carolina eggs were reburied. This technique was utilized after previous attempts by the conservationists. When eggs were moved within a few days following deposition, this resulted in very low hatching success percentages. This program has yet to be determined successful in terms of having increased loggerhead turtle nesting use of the Virginia beaches.

As discussed, embryonic development is already underway by the time loggerhead turtle eggs are dropped into their incubation chamber; however, an embryo is not visible to the naked eye until about eight days into the developmental process. The series of photographs on pages 104, 105, 106, and 107 graphically show the systematic changes which the contents of *Caretta* eggs undergo during the embryological developmental term. I have selected six day intervals to illustrate these growth levels.

As development of each viable egg accelerates in the last twenty days of term, metabolic heating, a rise in temperature generated by the increased volume of blood flow within individual embryos, radiates uniformly through the clutch. Eggs within the core of the egg column are thus equalized in temperature to that of those eggs at the outer circumference or top and bottom of the clutch. This metabolic heat within an egg complement has been measured up to six degrees F above that of the ambient sand temperature. A few days prior to actual pipping of the egg by prehatchlings, this temporary rise in temperature decreases until it becomes balanced to that of the adjacent beach soil. It is this phenomenon that permits parallel development and near simultaneous pipping of the egg shell by hatchlings at developmental stage thirty. This concurrent hatching is essential for survival of the group. The frenzied digging and later united nest exodus, are mechanisms which ensure that the turtles reach the ground surface together. Some sea turtle biologists have reported a secondary emergence, or a second group of hatchlings emerging the night following the first. Although this may occur on some beaches, I cannot recall ever having observed this on subregional beaches, nor do I have any records to indicate such. On Sanibel, specifically, the vacating of the nest chamber appears to be a one time effort, with the exception of a few hatchlings detained by some obstacle in their path or those that are delayed in their development for physiological reasons.

Coincidental to the embryo's rapid growth in the last twenty days of development, its carapace forms in a curled configuration in response to the shell curvature. In the meantime, by stage twenty-five, a caruncle, or egg tooth as it is commonly called, has developed on the anterior portion of the embryo's snout. When the magic moment of full-term is reached the now very active prehatchling is straining to escape and perforates the parchment-like shell wall with the caruncle, and pips or breaks through to its dark subterranean world. As pipping occurs, the hatchling, now working

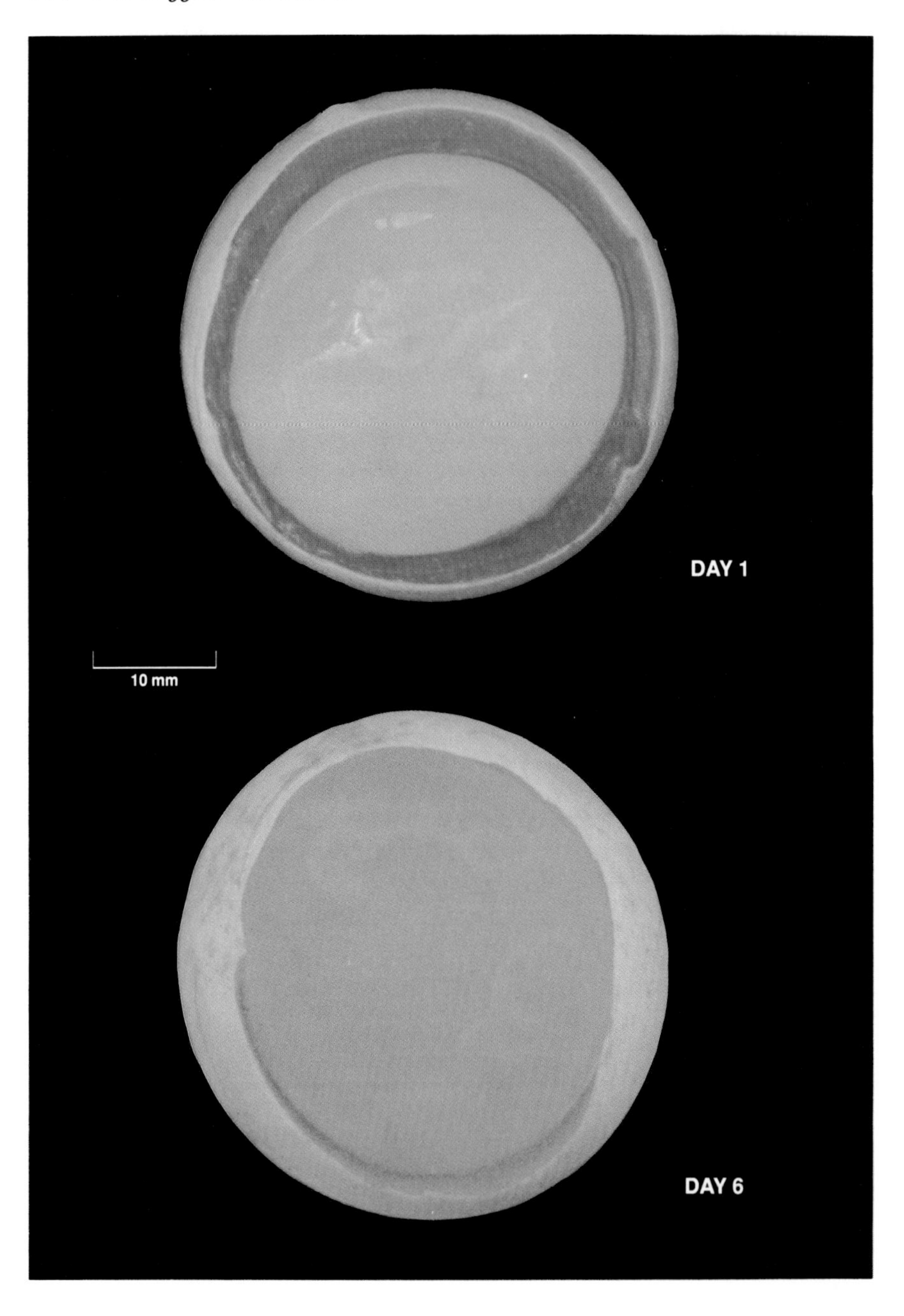

Plate 22. Days one and six of the developmental process. Day one represents stage six of the developmental schedule and day six approximately stage twelve. All photographs in this series were taken by Patrick D. Hagan.

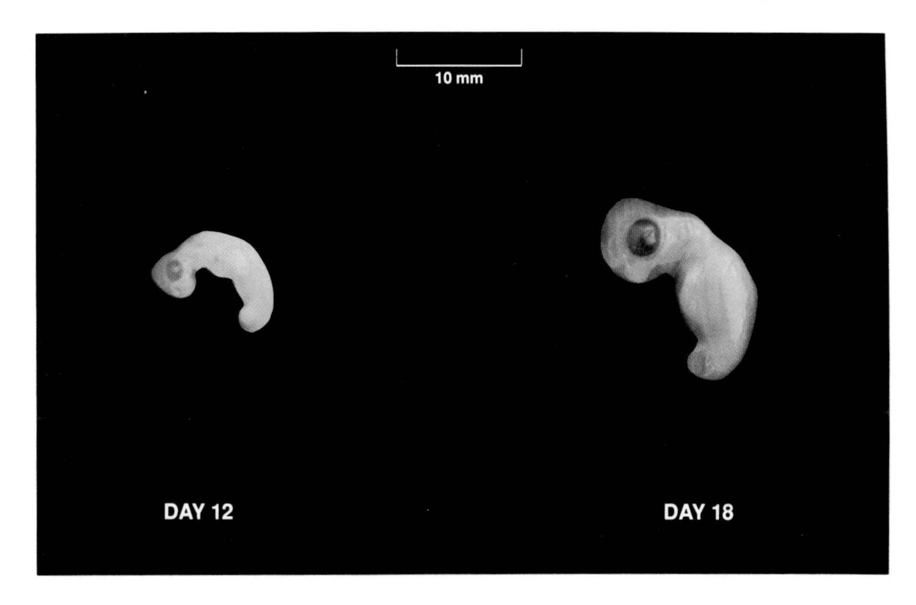

Plate 23. Day twelve (stage twenty-one). The embryo's head and eyes are present. By day eighteen (stage twenty-two) of development the carapace is discernible.

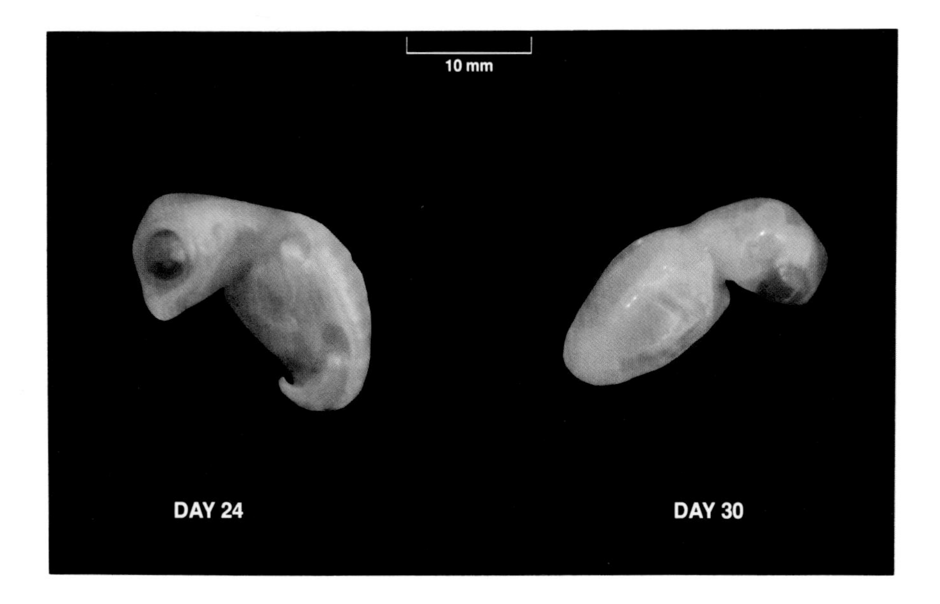

Plate 24. Day twenty-four (stage twenty-three). The foreflippers have developed. By day thirty (stage twenty-four) the carapacial scutes have formed.

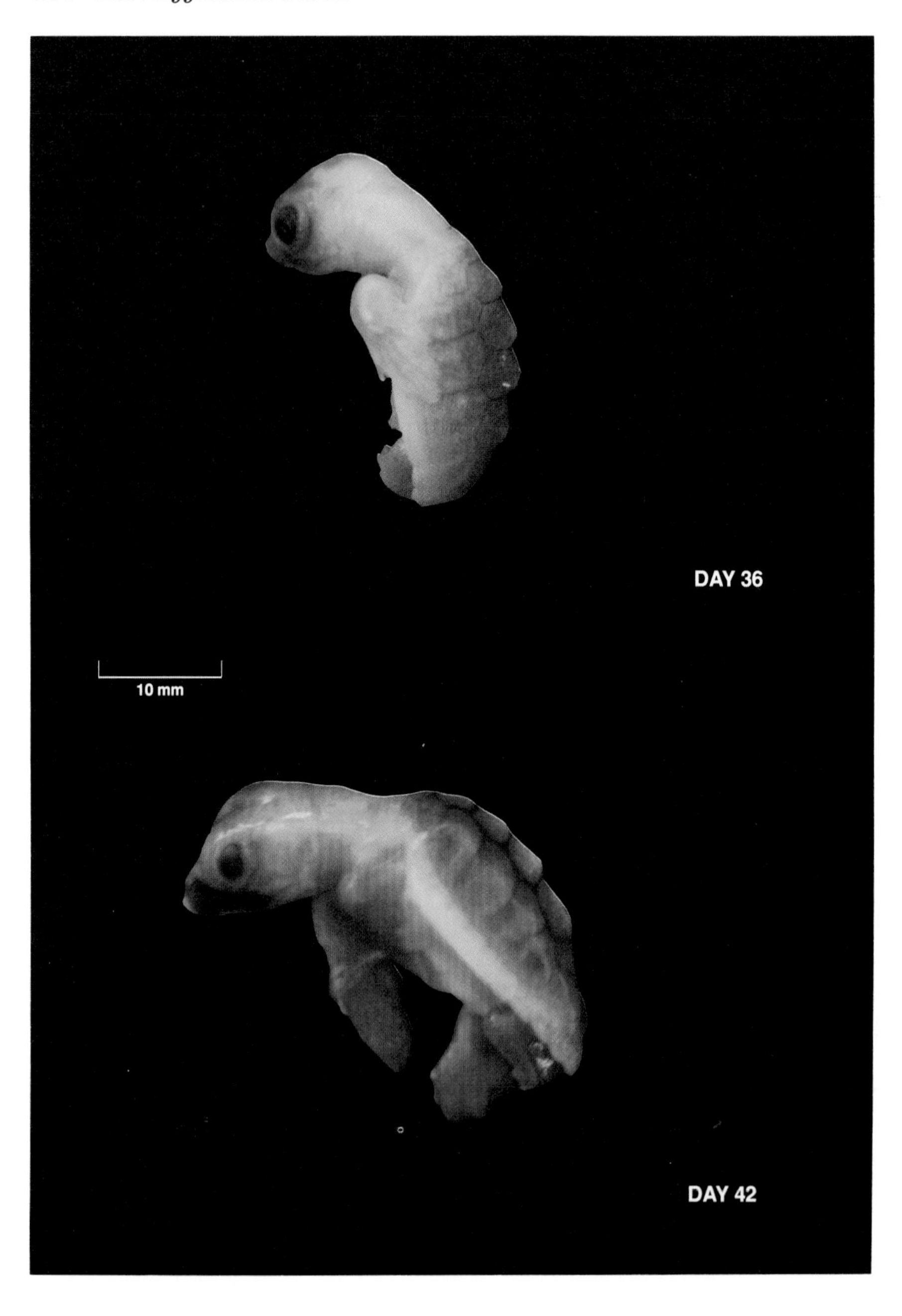

Plate 25. Day thirty-six (stage twenty-five). Claws are present on the flippers and pigmentation has started. By day forty-two (stage twenty-six) sexual determination has been attained.

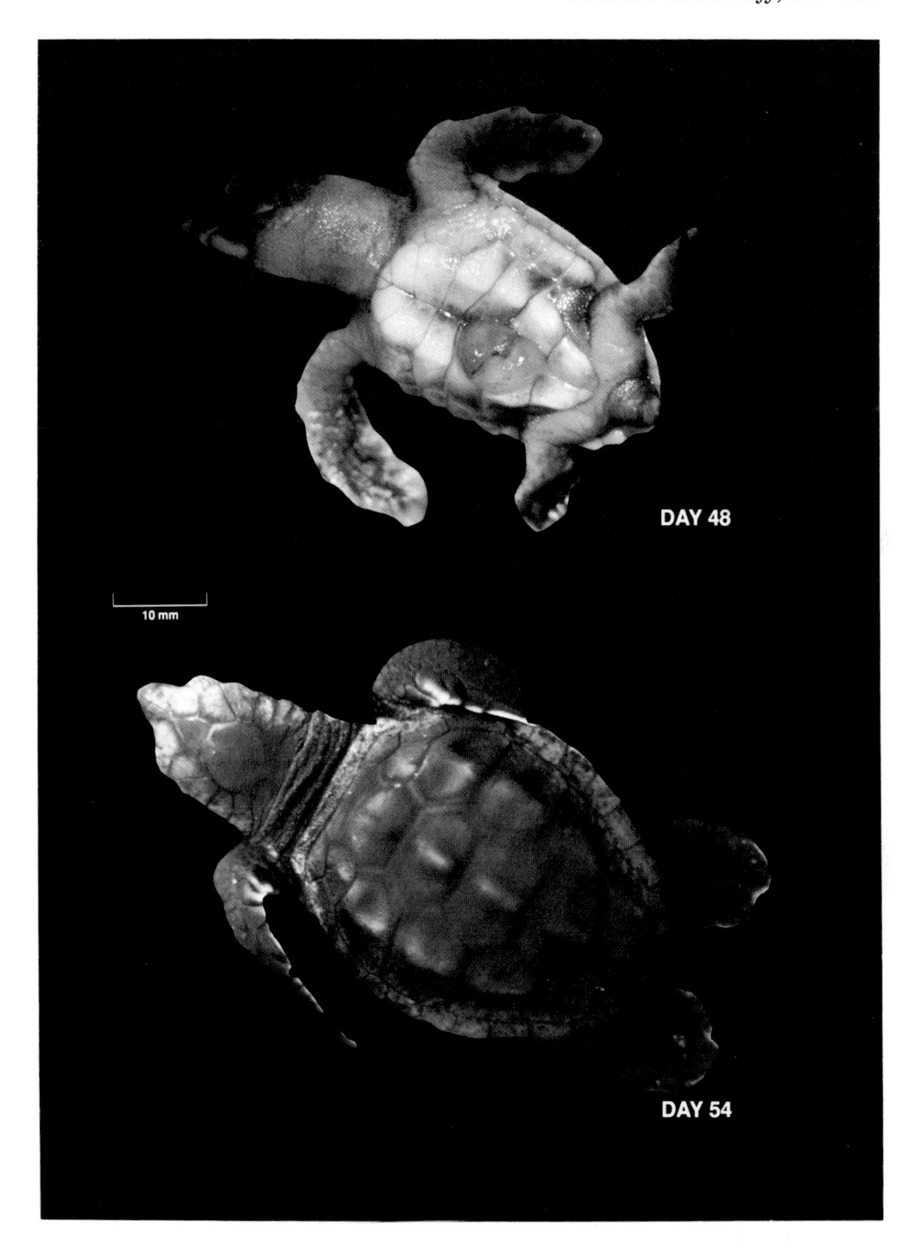

Plate 26. Day forty-eight (stage twenty-eight). The volume ratio of embryo to yolk is about equal. Day fifty-four (stage thirty) is the pipping stage. Much of the yolk has been absorbed but some is partially extended from the abdominal wall. Eggs prior to this stage were sacrificed to document the sequence of embryological development; however, this hatchling was nurtured and released.

free of the cramped, carapace-curving capsule, begins to unfold and straighten.

The allantois usually is severed during the hours while the late-stage embryo is frantically moving about inside the egg trying to break free. At pipping, and when straightening of the shell begins, a definite fold is evident across the plastron. Posterior to this cleft is a variable-sized external yolk sac from which the remains of the allantois streams. During the straightening process which occurs during the trip up from the confines of the nest hollow, this yolk sac is usually absorbed into the turtle's body cavity. By the time a hatchling loggerhead reaches the beach surface, the remnants of the allantois will have dried and fallen away.

In the past two decades, research of embryological development of turtles as a group, has produced startling new bits and pieces of information. The development of sea turtle eggs, including those of *Caretta,* were included in these pioneering investigations. This work has substantiated that *Caretta* lacks sex chromosomes and that incubation temperatures relate directly to the sex determination of hatchlings. Eggs which were incubated at 30 degrees C produced a combination of males and females, those eggs incubated above 30 degrees C resulted in only females being produced, and for the series of eggs which were incubated below 30 degrees C only males resulted. During the study which confirmed this strange relationship between temperature and sex determination, it was noted that development and successful hatching occurred between 26 and 34 degrees C. The sexual determination of a hatchling occurs between the twentieth and fortieth day, or within the middle trimester of embryological development.

The study which provided this new knowledge was conducted under laboratory conditions. Since there are no known external characteristics with which to distinguish the sex of little loggerheads, hatchlings were sacrificed and their gonads were histologically sectioned and examined microscopically. More recently, blood sampling to determine testosterone titer and even laparoscopy has been included in the growing technology being utilized to better comprehend all the implications of the role temperature has to development and the unique determination of sexuality. Little loggerheads no longer have to be sacrificed to determine sex ratios.

One of the ramifications of the discovery that developmental temperature influences the outcome of hatchling sex determination is the adverse impact that improper incubation temperatures may have on egg relocation efforts. For decades, conservationists have used artificial incubation containers, i.e., plastic buckets or styrofoam coolers, to house entire clutches during the incubation term. Some of these workers have simply collected eggs and placed them in sand-filled buckets, stored them, and occasionally sprinkled them with water to maintain a level of normal environmental requirements in an effort to approximate natural conditions. Others placed eggs in small styrofoam coolers and similarly maintained them. If the so called pivotal temperature which is somewhere close to 30 degrees C is altered because of such handling, then the possibility of skewing the sex ratio of relocated and artificially incubated clutches certainly is a paramount question to be considered if such practices are to continue.

The impact of *Casuarina* shadows on egg incubation temperatures and on the normal sex ratios of developing embryos are of genuine concern. Where these trees shade the beach it has been shown that soil temperatures can be reduced, and may be a degree or two cooler than open beach surfaces which are more exposed to the sun's rays. Most of the investigations on this important element of incubation have been conducted on the Atlantic coast. There, the unshadowed beach exists from dawn until about midday, or understandably the cooler part of the day. In the afternoon the tree shadows lengthen and shade the beach. On the Gulf coast, *Casuarina* shading occurs until midday, but in the afternoon the beach surface receives the full heating benefits of direct sunlight until sundown. I have examined surface and subterranean beach temperatures at a depth of thirty centimeters on open beaches and nearby beach systems that are shaded by Australian pines in the morning and unshadowed when sun heat is at its highest temperature in the afternoon. In 1988, I intermittently monitored sand temperatures on Sanibel Island during August and September and found that the temporarily shaded morning beach is less than one degree C cooler than the adjacent unshadowed beach. Later in the evening, by 1900 hours, the readings at the two temperature stations were almost identical. From these preliminary findings, I have concluded that the range in shadowed beach temperatures is not as great as those reported from the Southeastern shore of Florida and probably not as detrimental to hatchling sex determination. Large high-rise buildings, which have been inappropriately located on some prime loggerhead nesting beaches, are probably doing as much harm to natural sex ratios because of their shadows than those of the Botany Bay oaks.

In some circumstances, especially where nest sites are located well under the canopy fringe of the Australian pines and, consequently, subjected to prolonged periods of shade, the incubation timetables are extended because of cooler soil temperatures by several days. The normal incubation period of fifty-six days can be lengthened because of shading up to twelve or more days. These trees are not solely responsible for this extension. Some native vegetation, such as sea grape, can cause just as much shadow effect and torrential rains can swiftly reduce ambient sand temperatures, temporarily cool the egg mass temperature, and delay full term development. More field studies are required on turtle nesting beaches in Southern Florida to better comprehend the shading and sex determination relationship questions.

Over the past thirty-one years, I have relocated or transplanted *Caretta* egg groups into the natural beach face, with two exceptions. Once, we cut away the front of a styrofoam cooler and inserted a glass panel, filled the container with beach sand, and buried a loggerhead clutch against the glass. The section of foam that had been removed was repositioned and only removed periodically to examine the eggs and to observe and film the pipping and emergence of the hatchlings. Another season, I experimented with containers which were buried in the beach. Five gallon plastic buckets were utilized for this purpose. A series of a dozen or so half inch holes were drilled in the bottom and sides of the containers to permit water drainage, and beach sand was then placed in the buckets. When a clutch of loggerhead eggs were moved from their natural cavity a duplicate hole was dug in a bucket, the eggs were placed in this container and covered. Finally, a hole was dug in the hatchery compound and the entire bucket was buried just below the surface of the topsoil. The purpose was to allow removal and transportation of the clutches to a safe temporary site in the event of storm-related high tides. At the time I developed this methodology, I decided to phase out hatchery compounds. Mass transplanting of most of the loggerhead turtle eggs deposited on the Sanibel Island beach could no longer be justified to mitigate raccoon predation. When raccoons abandoned their natural prey base and opted for the contents of garbage cans, the major life-threatening pressure on turtle eggs was drastically reduced, and remained that way for several years. Today, I relocate clutches that are critically situated; i.e., too close to the water, to a safer site a few feet away and higher on the beach.

Since normal hatchlings typically vacate their developmental chamber as a unit, they leave behind weaker aberrant siblings and unhatched eggs. In nature, the turtles which are left behind usually remain in the nest, unable to pip the egg because of an undeveloped caruncle or simply too weak to dig out on their own. In this way, physically impaired hatchlings usually die inside the nest, rarely mingling later in the gene pool of the species. This is not always the case, however, because a leucistic loggerhead has been observed nesting on an Australian beach. I have examined a disfigured, genetically defective adult female which had stranded on Sanibel Island. Such specimens should have succumbed deep inside their egg cavity since only the fittest of a species are supposed to survive in nature.

One of the most frequently observed atypical conditions is abnormal scute arrangement on the carapace. If a hatchling has developed a scute pattern in excess of the normal number, they are known as supernumerary scutes. This form of deviation usually is present in hatchlings trapped in the nest cavity, hatchlings that are just pipping after normal sibling emergence, and late developing unhatched eggs. In the subregion, the latter may not pip for up to nine days after the generalized emergence and are doomed

Robert Pond

Plate 27. A normally pigmented loggerhead in the company of several white *Caretta* which were produced on Manasota Key, Sarasota County, Florida.

if researchers did not check the nest in unmonitored nest situations. It is these delayed term hatchlings which have the widest range of anomalous physical configurations.

Color variations are common within a group of hatchlings, but the most divergent colors occur in the delayed term hatchlings. Over the years, my associates or I have removed twenty-seven live white turtles from loggerhead nests in Southwest Florida. None of these were true albinos, for they all lacked pink eyes. These leucistic turtles almost consistently had carapaces with supernumerary scute alignments. One of these white loggerheads was maintained on Sanibel Island in a bait shrimp tank by Esperanza Woodring, until it died from unknown causes at the age of twenty-two months.

Some delayed term hatchlings have grotesque physical impairments. I have observed one double-headed loggerhead hatchling, and others with missing flippers, a few without eyes, and others with grossly misshapen

Patrick D. Hagan

Plate 28. Hatchling loggerheads, on reaching the beach surface, head straight for the water when conditions at the nest site are ideal.

mouth parts. Twice, I have opened late term eggs and found dead twin embryos, and in both cases these were completely separated prehatchlings with one common yolk. It is possible that some twinning may result in successful pipping and nest emergence, although no one has ever documented that this does occur. Successful twinning may account for small hatchlings, or runts, which are infrequently found in a group of normal sized emerging hatchlings.

Once the hatchlings break through the wall of their developmental capsule, the arduous task of escape begins. Tunneling toward the surface, the squirming mass of young turtles scrape away the ceiling and walls of the egg chamber. Periods of calm and rest alternate with frenzied spurts of upward digging. Soil that falls from the ceiling or walls filters through the struggling group and becomes the floor. Slowly, the hatchlings are transported in elevator fashion ever closer to freedom.

Beach surface soil temperatures are extremely hot in the summertime along the coast of the subregion, and solar radiation is the key factor in the timing of emergence by hatchling sea turtles. When the leaders of the group of hatchlings attain the soil strata level of hot dry sand that extends down three or four inches from the surface, they pause and the group rests. Once the upper beach soil cools after darkness falls or because of rainfall, the hatchlings erupt at the surface. Generally, emergence from the sand occurs within just a few hours following sunset.

Free of the nest, the turtles orient their route of travel toward the seaward horizon which is typically brighter than the vegetated land horizon. This phototactic response to light intensities guides the group to the surf and is a critical behavior element related to early survivorship.

16
Egg and Hatchling Mortality

Warren E. Boutchia

Once the eggs are deposited and covered by a nesting loggerhead and she turns and re-enters the marine environment, the eggs and any future hatchlings face a variety of predatory and mortality factors. When one considers the many predators and other life threatening elements which interact with eggs and pre-swimming hatchlings, the odds against their survival seem overwhelming, and long-term survival and future maintenance of the overall population seem doubtful. Yet, the species appears to be continuing despite the obstacles which predator population imbalances and other factors present. There have been many estimates of long-term survivorship for the group. The most common number, which is an educated guess, is that less than one percent of a nesting population's successful water-reaching neonates will survive, attain maturation, and be recruited into a nesting population. In this chapter, I will discuss the many negative aspects that have been imposed on the success of loggerhead turtle eggs and hatchlings.

Since time immemorial man has harvested sea turtle eggs for subsistence. Today, the practice continues not only in developing nations but within the United States and its territories. Despite protective legislation and enforcement, eggs continue to be harvested in this country from beaches where sea turtles come ashore to nest. Before sea turtles and their eggs became protected, it was a common summertime ritual for people to visit nesting beaches and collect sea turtle eggs in Florida. They are still held in high esteem by various cultures within our society. Loggerhead turtle eggs are eaten raw, pickled, or are utilized in the preparation of pastry recipes. When they are fried, the yolk will cook but the white does not congeal and, thus, is not considered very appetizing in appearance. When the yolks and albumen are included in pastries, they add richness and the white provides after-baking moisture which many users claim make turtle eggs superior to those of poultry.

Enforcement efforts by Special Agents of the U.S. Fish and Wildlife Service, Officers of the Florida Marine Patrol and the Florida Game and Fresh Water Fish Commission, have curtailed most of the illicit taking of eggs. Periodically, people are apprehended while collecting or transporting eggs but, despite these arrests, a black market trade in loggerhead turtle eggs continues to exist in the major cities along the eastern seaboard. At such places, archaic sociocultural traditionists continue to believe the erroneous notion that outstanding aphrodisiacal power lies within the thin wall of a sea turtle egg. In the late 1980's, black marketed loggerhead turtle eggs were selling in the neighborhood of $3.00 each.

Other than man, natural predators of loggerhead eggs include a variety of wildlife forms.

The Florida black bear, *Ursus americanus floridanus*, is no longer observed on the beaches of the subregion, but until the late 1950's, an occasional bear was seen near the Collier County coast close to Belle Meade, an old settlement between Naples and Marco Island, and near Royal Palm Hammock, a little further to the east. Historically, bears made seasonal visits to the Naples area beaches to feed on loggerhead turtle eggs. In the spring of 1958, a black bear was killed very close to downtown Naples by a City Police Officer with assistance from a rifle-toting private citizen. I was present when the animal was shot as it began to come down a pine tree which it had climbed for safety after being chased by hundreds of people. I have always conjectured that this poor bear had visited the Gulf beach looking for turtle eggs, and was only trying to get back to the Big Cypress Swamp where the species still occurs in small numbers.

The Florida wild hog, *Sus scrofa domesticus*, originated from escaped domestic pigs that have formed feral populations. Pigs are notable threats to sea turtle eggs wherever the two occur together. Within or near the subregion, feral hog populations occur on St. Vincent Island on the north Florida coast and Cayo Costa Island on the southwestern shoreline. During the summer of 1973, I studied the interaction between feral hogs and developing loggerhead eggs for six weeks on Cayo Costa. I found that the pigs usually located egg clutches following nest cavity discovery by raccoons. Pig-excavated nests were easily identified because of the sizeable depressions which the animals typically rooted out to locate and eat the eggs. These craters were examined and, in all instances, no eggs remained.

Feral pigs should not be allowed to remain as part of the resident wildlife population of barrier islands that are utilized as nesting beaches by sea turtles. Agencies which have stewardship of lands where such negative interaction occurs between species, when one form is threatened by a non-native animal, must take serious steps to intervene and remove the animal jeopardizing the threatened species—in this case, pigs.

The gray fox, *Urocyon cinereoargenteus*, is uncommon, but occasionally has been observed, by coworkers and me on Collier County beaches excavating loggerhead turtle nests and eating eggs. On the same beaches, their tracks have also been observed around the vicinity of recently plundered nests. Because of their low population levels the gray fox is not a major predator of sea turtle eggs in the subregion.

Of all the terrestrial predators found within the subregion, the raccoon, *Procyon lotor* ssp., is by far the most damaging to the successful production of loggerhead turtles. Established resident populations of these furbearers exist on all the subregional barrier islands. In some areas, individual and family groups of raccoons migrate to the outer beaches during sea turtle nesting time, i.e., at Cape Sable beaches in Everglades National Park.

Raccoons affect loggerhead turtle success during two stages of the reptile's life history. These periods are: as developing eggs within the subterranean cavity and, later, during the postemergent hatchling stage as little loggerheads dash to the water. On the beaches of Southwest Florida with very high raccoon populations, where the mammals rely heavily on the turtle's eggs for nourishment, most nests are opened within one or two nights following deposition. Sea turtle workers on Florida's east coast report that in some areas, raccoons do not predate nests until the end of incubation. On those beaches the animals locate nests via olfaction, responding to the odor generated by pipping eggs. In the thousands of loggerhead turtle nests which I have closely monitored, both natural site clutches and those that were relocated, raccoons have invaded the nests within a night or two of deposition, or later if they accidentally encountered the hatchling column emerging from a nest and crossing the beach.

Raccoon predation is easily identified. Broken egg shells are scattered in the vicinity and 'coon tracks are everywhere around the nest. Unlike the total devastation of a nest which has been excavated by feral pigs, raccoon penetration usually involves removal of sand until the top layers of eggs are exposed. The 'coons then use their ambidextrous forefeet to remove the eggs. Consumption of most of the eggs is customary. In cases where some eggs survive the initial opening, the outlook for the remaining eggs is not good. I have occasionally removed surviving eggs from predated nests, carefully cleaned them of egg fluids, marked the shell surface with a pencil

Edward J. Phillips

Plate 29. Raccoons are the greatest threat to the success of loggerhead turtle nests on most subregional nesting beaches.

for later identification, and reburied them in some other nest that was deposited on or about the same night. In these limited cases, survivorship of the relocated eggs was within normal ranges. The few eggs remaining in a predated nest, should they survive the threat of repeat raccoon visits and successfully develop, become hatchlings which do not have the numerical forces required to escape the egg chamber.

In instances where a clutch of eggs escapes early discovery by raccoons, predation may still occur. When a group of hatchlings reaches the surface of the beach and begins the traverse to the water, they are normally well spread out between the nest and the edge of the surf. If raccoons should encounter them at this critical period when they are exposed on the beach, many never reach the water. Some hatchlings are eaten completely, others are partially eaten at the yolk sac remnant on the plastron; all are decapitated, and only a small percentage may reach the relative safety of the surf. I have seen clear evidence of this scenario occurring many times on Sanibel Island over the years.

On unbridged islands, such as Cayo Costa, the Cape Romano group, and Cape Sable, loss of egg complements is very high because of raccoons.

On the Cape Romano outer beaches, destruction of loggerhead eggs is 100 percent each summer. Bridged islands with human residential and resort development have varying degrees of loss. Sanibel Island was experiencing approximately a 75 percent annual loss of loggerhead nests to raccoons until land development and the human population increased in the early 1970's. Sanibel's raccoons quickly learned to alter their prey base and depend on the abundance of waste food in trash cans and refuse dumpsters. This continued for almost two decades until the urbanized raccoon population rose to rampant levels and the supply of human garbage was insufficient for the dependant population's needs. By the summer of 1988, raccoons again were being observed on the beaches at night. Ten percent of our marked and monitored nests were destroyed by 'coons that year.

David Spicer, a graduate student from Iowa State University, was on Sanibel in 1988 conducting a study of the Island's raccoon population. He estimated that Sanibel Island supported one raccoon to every .5 acres of habitat, or somewhere around 5,000 raccoons. Raccoons were at an all time high—until Mother Nature, the great population regulator, said enough is enough. During the winter and spring of 1988-89, canine distemper swept through the raccoon population and decimated their numbers. The wildlife care organization, Care and Rehabilitation of Wildlife (CROW), based on Sanibel Island, picked up and euthanized over three hundred raccoons during the outbreak of this always fatal sickness. All animals were confirmed to have the disease. Dr. Marc Elie, the Executive Director of CROW, estimated that at least 90 percent of some very dense localized populations and 75 percent of the general population succumbed to the ailment. It will be interesting to observe any changes in the number of loggerhead turtle nests that are predated by 'coons in future seasons.

In the early 1960's, loggerhead nests on the sections of Sanibel Island beach having the highest concentrations of raccoons were fitted with wire covers for protection. As public use of the beaches by nonresidents increased, the protective devices were vandalized by uncaring people and I abandoned this practice in 1963. Various attempts were made to discover or develop a repellant formulation that would chemically camouflage egg sites and deter nest discovery and damage by raccoons. A wide variety of chemical techniques were employed including household ammonia, "Chaperone" dog repellant, urea granules, mothballs and human and dog urine. Some of these appeared to work, but were not reliable and none appeared to provide the level of protection I desired. During his loggerhead work on Bonita Beach, Jim Vanas insisted that application of human urine over and around an egg clutch was only effective if the urine contained a high level of Budweiser beer!

In 1969, I initiated a transplantation program on several islands along the Southwest Florida coast to relocate eggs which were threatened by raccoon populations. Eggs were moved into hatchery enclosures to safeguard them from raccoons. To help ensure egg viability, only those eggs less than twelve hours old were relocated. Others which were outside of this time threshold were not collected and remained on the beach where deposited; however, these were marked, or their locations logged, and monitored during incubation. On Collier County beaches, such as Keewaydin (Key Island) and Cape Romano, the hatchery concept was significantly important because loss of eggs to raccoons occurred without letup and these losses continued at a very high percentage. During the years in the 1970's, when these hatcheries were operational, they were very successful. Today, a loggerhead turtle hatchery is functional each summer on Keewaydin under the auspices of The Conservancy in Naples. I am pleased that I pioneered this work and contributed to raising public awareness of the plight of sea turtles in Collier County. The Sanibel Island hatchery program was phased out in the mid-1970's.

The opossum, *Didelphis virginiana,* is common in most of the mainland sections of the subregion, but it does not occur on all of the barrier islands. In fact, opossums were unknown on Sanibel Island before the Sanibel Causeway connected the Island to the mainland in 1963 making it accessible to nonresident mammals such as 'possums and bobcats. Elsewhere, these mammals are considered to be a minor predator of loggerhead turtle eggs. Opossums may only seek eggs from nests that have been opened by raccoons who left a few eggs behind.

The striped skunk, *Mephitis mephitis,* also has a limited distribution on the barrier island chain of the subregion. Although they are known to consume sea turtle eggs in the Southeastern United States, I have no records or other evidence of nests being opened by skunks in Southwest Florida. Skunk predation may be a secondary form, similar to the depredation by opossums.

The nine-banded armadillo, *Dasypus novemcinctus,* now occurs on many of the barrier islands in the subregion. In some instances, these animals were introduced by man, such as in the case of Captiva Island which had the animal introduced in the late 1950's by a resident. Later as their population succeeded, the animals spread to adjacent Sanibel Island. I have observed armadillos swimming in the middle of Pine Island Sound near these islands, so it was probably only a matter of time before they reached Sanibel-Captiva from the mainland on their own. Over the years, I have observed armadillos nosing about in the vicinity of loggerhead nests; however, I have never made any personal observations which would indicate that these strange mammals are more than a rare and minor predator of loggerhead turtle eggs.

Feral populations of the common cat, *Felis domesticus,* are widespread on the outer islands of the subregion. Both they and pet house cats which roam near beachfront residences at night are routinely encountered on the beaches. I have never observed a common cat excavating a loggerhead nest, but I have observed them pawing and carrying off hatchlings that were moving from their nest site to the Gulf of Mexico.

The Norway rat, *Rattus norvegicus*, and the roof rat, *Rattus rattus*, are frequently observed at night on the beaches of the subregion. The latter is far more common than the former, at least on Sanibel Island. I have no documentation available to indicate or suggest that these introduced rodents open loggerhead turtle egg cavities; however, they sometimes intercept postemergent hatchlings and prey upon them. Through the years, I occasionally have found partially eaten loggerhead hatchlings bearing rat incisor teeth indentations, next to recently vacated nest sites. Individual rats, on the rare occasion when they encounter an assemblage of emergent hatchlings, may remove one or two turtles from the group, but they do not wreak havoc like raccoons.

Emerging from their nest under the protection of darkness, hatchlings circumvent the dangers posed by most avian predators; however, this selected timing does not eliminate potential life-threatening hazards facing the hatchlings from all birds. There are three common species of nocturnal birds that are high risks, in my opinion, to the safe passage of newly emerged loggerhead turtles on their journey across the open beach.

The black-crowned night heron, *Nycticorax nycticorax*, is abundant on the Gulf beaches of Southwest Florida at night. During one night's first turtle patrol pass of 12.5 miles, from east to west, in June 1965, on Sanibel Island, I counted sixty-seven of these birds lining the edge of the surf. I have seen small groups of black-crowned night herons suspiciously close to sites where emergent hatchlings had crossed the beach just moments before I arrived on the scene. The black-crowned night heron feeds on fish and crabs and forages in the surf. In my opinion, these birds do indeed take and consume hatchling loggerhead turtles; however, I have never been able to substantiate this through personal observation.

The yellow-crowned night heron, *Nyctanassa violacea*, also frequents the Gulf beach zone at night, but is not as commonly observed as the former species. The yellow-crowned occurs on the Sanibel Gulf beach at a ratio of approximately one to twenty-five of the black-crowned night herons. Although I am personally convinced that this heron, too, will prey on hatchling loggerheads, I have never verified this fact.

On bright, moonlit nights along Southwest Florida's barrier beaches, great blue herons, *Ardea herodias*, are frequently observed wading in the surf. As in the case of the night herons, I have no verification that this species feeds on hatchling loggerheads, but their very presence at night on the beaches combined with their known appetite for a variety of organisms, certainly makes the predator-prey relationship between the two very likely.

The abundant fish crow, *Corvus ossifragus,* is a diurnal species that frequents the subregion and often occurs in large groups. In the rare event of hatchlings crossing the beach during daylight, crows quickly congregate and prey on them. Hatchlings which may have been attracted inland during the night because of the interference of artificial lighting on or behind the Gulf beach ridge and have escaped other imminent dangers, are doomed at sunup when crows find them. In 1969, while I was filming a movie on the life history of *Caretta,* I released an assembly of hatchlings on the beach. To my dismay, groups of fish crows were attempting to pick up little turtles and, despite my efforts, were successful in picking up and flying off with some hatchlings.

Seabirds of various species, such as the magnificent frigatebird, *Fregata magnificens,* are known to take neonate marine turtles at sea. A segment of a popular television series made in the 1970's, provided viewers with some rather dramatic documentation of this predator in action. However, since sea turtles typically emerge from their nests at night and frigatebirds are diurnal, there is seldom such interaction between the two on the beach as was staged to film this documentary.

I have received one report from a field worker of a black racer, *Coluber constrictor priapus,* being present in a loggerhead turtle egg cavity. While excavating a nest site, Fran Wright was surprised to find a live racer buried with the eggs. There was no visible indication of how the snake gained access. It could have reached the clutch via ghost crab tunnels which may have been situated close by, and led into the nest.

There are confirmed reports of various snakes and large lizard species around the world being sea turtle egg predators. Since there was no evidence that the black racer discussed above had eaten or broken any eggs, there remains a reasonable question of why the snake was present. It could have simply found refuge, or the racer could have been attracted by the eggs. An adult black racer certainly could consume loggerhead eggs or even hatchlings.

The ghost or sand crab, *Ocypode quadratus,* is a recognized international predator of both developing eggs and hatchlings. This crab constructs and inhabits burrows in the same beach zone in which sea turtles deposit their eggs, and over which hatchlings must rush to reach the sea. The crabs can tunnel directly into egg clutches and feed on the eggs underground, or remove them from the nest through the passageway. The openings which crabs construct that reach egg clutches, also permit the later entry of insects such as ants and flies, or even rats. Flies deposit their eggs on turtle eggs and their larvae can seriously damage a nest complement. Once eggs are opened inside the cavity, by whatever means, and a passageway leads to the beach surface, bacterial incursion can destroy the clutch.

In coastal areas where the ghost crab is numerous, they may intercept

Plate 30. The ghost crab inhabits the same beach zone utilized by *Caretta* for nesting habitat. These crabs prey on loggerhead turtle eggs and hatchlings, but not to the level of severity posed by ants and furbearers.

postemergent hatchlings, pull them inside their tunnel, and eat them. If enough crabs converge on an emergent group of turtles, several from the clutch can be lost. I have never developed any statistics on which to base a percentage of loss of loggerhead eggs to ghost crab predation. The interaction is subtle and generally goes unnoticed but, when known nest sites are examined just after hatchling emergence, I have found skid, or resistance marks, among the scores of tiny turtles tracks. These marks invariably have led away from the group's route of travel and ended at the entrance of a ghost crab burrow.

The native Florida fire ant, *Solenopsis geminata,* has frequently been observed as an invader of loggerhead turtle egg clutches on Sanibel Island. It appears that these ants cause insignificant levels of damage to prehatched egg groups. It is during the critical time when hatchlings have left the egg and are ascending from their developmental chamber that they are most vulnerable to ant attack.

Should a clutch be deposited in a partially shaded or vegetated location where these ants forage, the developing turtles are seriously threatened; however, unshadowed sites are not totally excluded from this threat. During emergence activities, the ground above the hatchlings often cracks through to the surface, and the open cracks offer an avenue of access which the ants will follow to the hatchlings. Once passage to the turtles is opened, it is but a brief time until all hatchlings fall victim to the formic acid injected at the time of the invading insect's fiery sting. An entire group of loggerhead hatchlings may be killed before they can reach the beach surface. On Sanibel Island, fire ants presently are the most dangerous predator that pre-emergent neonate loggerheads must contend with.

To compensate for this serious problem and avoid fire ant depredation, we inspect each nest on Sanibel Island as it approaches full term as part of our monitoring program. Field workers carefully examine the vicinity of a developing nest and, when fire ant hills are found, or other evidence of ant activity is present, they are treated with Amdro, a commercially available pesticide formulated to kill these ants. The compound is applied to the beach surface far enough away so as not to leach through the ground and reach the eggs. This program has proven to be very successful in combating this major cause of hatchling mortality.

It is common for packed sand particles to contribute to clutch or pre-emergent hatchling mortality, and this phenomenon has been well demonstrated on Sanibel Island. When soil samples from the entire length of Sanibel's Gulf shoreline are visually examined, it is evident that the beach sand becomes progressively coarser as random soil collections are inspected from stations leading to the west and away from the Sanibel Lighthouse.

Beach soil types on Sanibel Island appear to be distributed according to wave dynamics. East of Knapp's Point, the southernmost projection of the Island, very fine sand, known locally as sugar sand, is the predominant soil. West and north of Knapp's Point, the soil consists of a high percentage of coarse shell fragments. Appropriately, the western portion is moderate energy; whereas, the eastern section is considered a low energy beach system.

During the almost daily torrential rains in August and early September when it is not uncommon for two or three inches of precipitation to fall in an hour's time, developing clutches are critically jeopardized. Depressions in the beach berm and foredune systems between the Lighthouse and Knapp's Point are flooded during these intense rains and remain so for hours, because of the impermeable nature of the soil's particulate fines. Eggs situated in such locations on eastern Sanibel are killed because of long-term flooding. Flooding prevents gas exchange or cools the clutch temperature to a level which arrests embryological development.

Many in-nest groups of dead hatchlings have been examined during the monitoring program which is designed to determine overall nesting success. Rain-damaged sites have consistently been discovered in the eastern

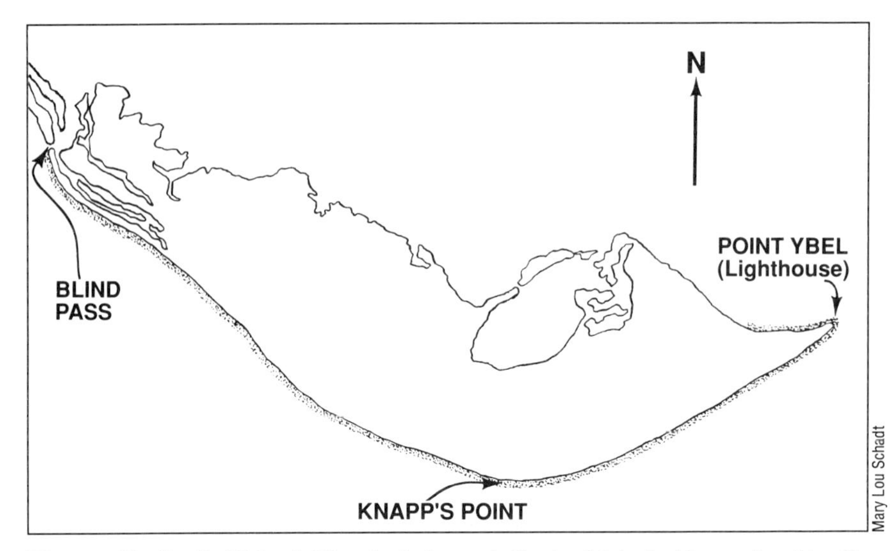

Figure 40. Sanibel Island. The shaded area indicates historical loggerhead turtle nesting beach.

impermeable section of Sanibel, while very few rain-caused nest mortalities have ever been observed in the coarse, well drained, western section of the Island. A combination of precipitation coincidental to tidal surges created by tropical storms or hurricanes is another matter. Regardless of the ability of beach soil to drain rapidly when high, flooding tides inundate nest sites, drowning of eggs and hatchlings can occur. During the hatchlings' journey from egg to beach surface, they are highly vulnerable to inundation and the flooding discussed above. When an impermeable egg cavity floods and remains flooded because the water simply cannot drain off the site fast enough, egg or pre-emergent hatchling mortality is one hundred percent.

As a management strategy to enhance survival of those egg groups which are deposited on the impermeable eastern Sanibel beach, we now excavate the nests the morning following deposition. They are transplanted to the higher, better drained beach at the western end of the Island and monitored through their incubation.

Many beaches within the subregion are extremely dynamic and subject to the forces of severe erosion; for example, over the last eighty-five years Captiva Island has become over a thousand feet narrower along its Gulf exposure. Many geological and hydraulic factors are responsible for such a dramatic erosion rate. Upland property owners, in cases where they elected to build too close to the beach or because of accelerated erosion like Captiva's problem, see only one technique as their salvation—beach nourishment.

Beach nourishment is the excavation and transfer of usually submerged soils to the diminishing beaches. The transferred soils may originate from navigable channel maintenance or from offshore shoals selected as spoil source sites. Typically, such work requires the use of a hydraulic dredge which utilizes water as the medium with which to transport soils from the spoil site to the upland that is being rebuilt or extended. The saturated soils can pack extremely tightly as the waterborne fragments are deposited on the beach face.

There has been some speculation and cursory studies conducted on the success of such improvements to natural beaches and the disadvantages of such restoration projects. Some investigators are of the opinion that the new water-packed sands added to nesting beaches are not conducive to improvement of sea turtle nesting habitat. Most field work to evaluate turtle use and nesting success on nourished beaches has been conducted on the Florida east coast. Atlantic coast beach soils are different in composition than those of Southwest Florida, having coarser fines and a higher percentage of silica sand. The biogenic soils found on many beaches of the subregion apparently do not react to water loading by rain and tide in the same fashion.

The dredged spoil can create a hardened beach surface that makes it difficult for nesting turtles to excavate an egg cavity. On the Atlantic coast, nourished beaches are best disked, or scarified, with heavy machinery to reduce this problem and create a beach surface texture more in keeping with the beach consistency prior to deposition of spoil. In 1984, ten thousand feet of the Captiva Beach were renourished south of Redfish Pass. We carefully monitored the beach which had not been disked or otherwise surface disturbed the next nesting season, and found a higher density in turtle use than before nourishment. The success of all loggerhead nests which were watched during the summer were all within the normal range of productivity for the subregional nesting population.

The real threat to sea turtles, because of dredged nourishment projects, comes not from soil compaction, but from unregulated covering of nest sites with an inordinate amount of additional topsoil. To mitigate the potential damage to sea turtle eggs from dredging projects that start during the sea turtle nesting and hatching seasons, the United States Army Corps of Engineers and appropriate State agencies, as part of their permitting or contracting processes, include provisions for beach monitoring and relocation of any nests located within the boundary of the spoil deposition site. By daily, early morning examination of beaches, clutches of turtle eggs which were deposited the prior night can be safely excavated and moved out of the construction area without reducing their viability. In 1988, I entered into an agreement with the Captiva Erosion Prevention District to conduct such a transplantation and monitoring program. Captiva eggs were excavated the morning following deposition and relocated to a natural beach site on nearby Sanibel.

In 1989, the Captiva beach was examined by a representative from the Army Corps of Engineers and me after completion of the new beach. This was done to evaluate the degree of compaction of the beach soils as a result of the nourishment, and to determine what impact the tightness of the beach might have on nesting loggerheads, resultant egg development, and hatchling survival. The beach was examined visually, by hand with a shovel, and with a penetrometer, an instrument used in assessing beach compaction. We concurred that the new beach was very similar to what occurred naturally on nearby Sanibel, so the Captiva Erosion Prevention District was not required to scarify the Captiva beach. In 1989, the Captiva beach was carefully monitored for loggerhead utilization and hatching success rates—both proved to be normal for the island.

There are incidental problems associated with dredging operations that may negatively affect sea turtles. The water turbidity generated by dredging intake and dispersal and general disturbance because of the work in the close proximity to an active nesting beach, may cause unknown reactions by sea turtles during their internesting intervals. What impact this will have on an individual turtle's future reproductive visits is unknown at present.

Many of the beach areas which are feeling the boom of development are falling under the spell of a new breed of resident—the beach manicurist. These are recently-arrived people from some other part of the country who are out of their lifelong urban element. It is common for them to insist that natural beach debris, such as parchment worm *(Chaetopterus veriopedatus)* cases, Gulfweed and other marine plants, have no place in front of their condominium development. Managers of such units hire mechanized equipment that rakes or scrapes away the "undesirable" organic material. Increasingly, Gulf frontage in Southwest Florida is becoming influenced by this "Miami Beach Syndrome." Shortly after the City of Sanibel was created in 1974, the then strong environmentally oriented City Council passed an ordinance that prohibits mechanical clearing or grading of the beaches within the City limits which includes all of Sanibel Island. I was a member of that first City Council and party to many such actions which were designed to protect the ambience and environment of Sanibel. I see some of these essential qualities of life slowly eroding because of attitudinal changes in the City's leadership as Council members come and go, and I wonder just how long it will be before the Sanibel beach I love will be just another groomed, smooth, beach.

Grooming of beaches with scraping or raking implements during the time of year that sea turtle eggs are buried just inches beneath the surface can result in clutch mortality. Removal of too much topsoil above developing eggs can lead to dehydration of the upper eggs or break eggs at the top level. Broken eggs can attract predators to the clutch location, or the nest can be bacterially invaded, both factors resulting in severe reduction, or complete destruction, of a nest's productivity.

Another problem, related to public use of the beaches, has recently surfaced on Sanibel Island. Many of the resorts provide beach furniture for use by their guests. Some of these items are fabricated from very heavy pressure treated lumber. On the night of June 6, 1989, we stopped in front of a condominium to investigate a crawl, which had been made the previous night after we left the beach. At the turtle-disturbed nest site we found a heavy beach lounge, with one of its two by four inch wooden legs imbedded in the sand, directly over where the eggs should be. We moved the lounge, dug away the topsoil, discovered seven broken eggs, and found the upper levels of eggs to be totally infested with maggots. With her bare hands, Kathy Boone carefully removed all the fly larva, and we covered the site with fresh moist beach sand, raising the elevation somewhat to better protect the eggs should the lounge end up back in its original position. The nest was marked and revisited for scheduled monitoring of its fate on August 5. We found that the nest was successful, despite the early damage, and produced seventy-three hatchlings from a total complement of one hundred one eggs. One dead neonate was found deep in the nest along with twenty addled eggs.

In some situations, developing eggs are threatened by a hidden enemy. In beach zones that are thickly vegetated by sea oats, *Uniola paniculata*, the roots of this plant, especially those that were earlier damaged during egg chamber excavation by a loggerhead, can seriously impact the outcome of successful incubation. During the rainy season, which peaks in July and August, growth of the plant's root system is accelerated. I have observed egg groups that were completely encapsulated by sea oat roots, that had grown profusely through the chamber, making it impossible for pipping hatchlings to escape the egg. In other cases, during the monitoring of full-term nests, we have often discovered hatchlings that were permanently snared by rootlets of this and other beach plants which had tightly encircled their heads or flippers. On beaches where there is no follow-up examination of nest success, root growth certainly accounts for a potentially high loss of hatchlings.

On some soft berm beaches, I have found desiccated hatchlings that had been trapped in the deep ruts caused by the operation of heavy, narrow-tired, motor vehicles on the beach. Tire imprints can create a series of steep-sided depressions and hatchlings which encounter these, fall into them enroute to the water. The sides of these ruts usually slough very easily and many of the young turtles may find it impossible to climb out and pass across a beach which has a number of sets of water-paralleling vehicle

Rickie Cooper

Plate 31. A hatchling ensnared by roots is trapped in the depression that usually forms at the egg cavity following emergence. Had its plight not been discovered during nest monitoring, ants, crows or heat would have destroyed this turtle.

tracks. Ants face a similar situation when caught in the cone-shaped sand trap of a doodle bug, or ant lion. Should the depressed obstructions be present in numbers sufficient to completely impede the passage of hatchlings, the entire group is trapped, unable to reach the water, and will soon fall victim to the summer sun's heat the next day—unless they are discovered and carried off by fish crows at dawn.

Our nest marking method involves driving a twenty-two inch long stake into the beach surface a varying distance upland from the eggs, unless vegetation would obscure it. If this would be the case, the stake is driven a short distance seaward of the site and its location noted on the field data sheet. We used reflective tape in a color-coded system to identify the site as to month and period within a month, for later monitoring. The colors utilized for each month are: May—red, June—white, July—orange, August—yellow, and if a nest should occur as late as September—a combination of red and white. For nests deposited between the first and fifteenth day of a month we encircle each stake with two pieces of tape. Nests

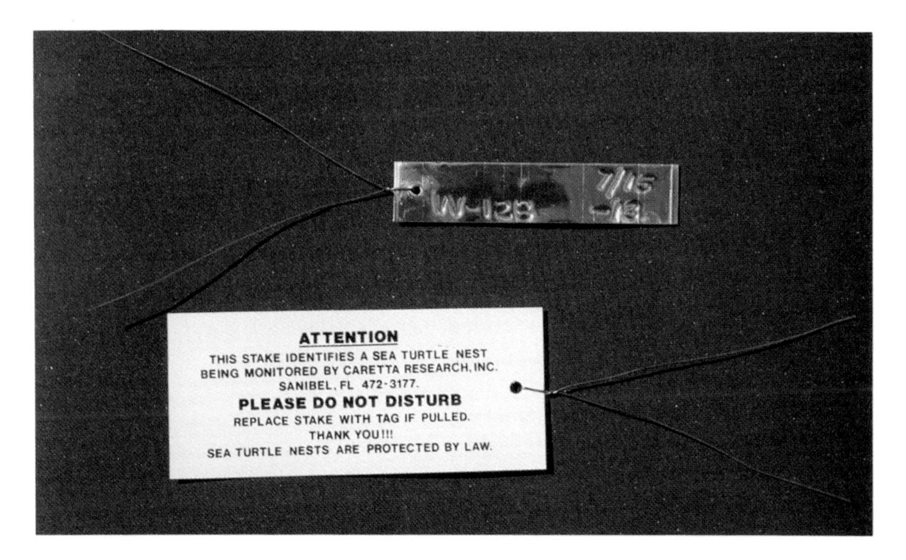

Plate 32. Two types of nest identification tags are used on Sanibel and Captiva Islands.

deposited from the sixteenth day through the last day of the month have three strips of tape affixed to them.

These identification stakes are located in the nesting zone which is higher on the beach than most pedestrians travel, and they are not easily detected during daylight. At night, they come alive with color and are easily located. In the early days, I used only red tape in various numerical codes, to divide the months. To travel the Sanibel Island beach in the summertime was like celebrating an early Christmas with all the red illumination bouncing from the stakes. I would preface my introductory remarks on loggerhead turtle biology, to those novice people who joined me for a night on turtle patrol, explaining they would have the unique privilege of visiting Sanibel's red light district! I would sometimes get a chuckle in response, but I would say no more on the subject—until we had made our first pass of the beach.

Two tags, which are pictured above, are applied to each stake. One is a small paper-backed aluminum tag; the other a printed vinyl tag. The clutch number, i.e., W-128, date of oviposition, and distance of the stake, in feet, from the actual nest, are handwritten on the aluminum tag. The larger vinyl tag bears a brief message identifying the stake as a nest marker, and it asks a finder not to disturb the stake. Both tags are covered with sand to make the marker less obvious, although in some areas of high public use we may leave them unconcealed to further reduce removal or vandalism of the stakes. We have found that if people see the tags before they pull the stake out, we can expect them to remain full term—even at entrances to the Island's most popular beach park. On remote sections, less attention is drawn to the stakes if the tags are buried. These stakes are seldom vandalized, we lose less than three percent of our nest stakes each season. We can sometimes locate the few nests where stakes are pulled out because we log the vehicle odometer readings down to a hundredth of a mile at each nest by subdividing tenths. Some nest site depressions can be located for monitoring examinations based on the odometer readings—if we have not experienced a storm that would alter the beach by covering sites with an overlay of new sand.

As hatchlings negotiate the terrain during their crossing of the beach face, moving from the nest site to the water, they may face serious setbacks other than predators.

Large, partially buried trees which are carried to the frontal beach by tides or storms, also present a danger to the water-seeking hatchlings. Sea turtles, generally, are not very good at backing up; in fact, the leatherback turtle doesn't. Should a large forking tree with the apex of the fork located towards the water be in the hatchling's avenue, the situation becomes very serious. Entire groups of hatchlings bent on going in one direction only will pile up at the tree crotch and exhaust themselves attempting to climb over the obstacle. In this scenario, too, an entire otherwise successful group of hatchlings can perish because of desiccation or discovery by the ever present marauding fish crows or raccoons.

On reaching the beach surface from the depths of their egg cavity, individual hatchlings visually examine their surroundings and respond to the prevailing intensity of illumination. Usually, this brief horizon scan, done individually, orients the hatchling group, attracting them to the bright seaward horizon, as opposed to the darker vegetated horizon of the upland. In a natural setting, this reaction to light guides the hatchlings to the edge of the surf. Along coastal areas which have been altered by development and construction, the seaward illumination is often overpowered by the bright lights of buildings and associated exterior accessory lighting, or even beach bonfires. Under these circumstances, hatchling sea turtles emerging from the nest mistake the visual cue and are misdirected inland, away from the water. This inaccurate reaction of the hatchlings to the artificial light source can lead to disaster.

Mass mortality of hundreds, or even thousands, of hatchling sea turtles has been reported in Florida because of the interaction between their ocean finding needs and the stronger opposing attraction of bright coastal lights. Young turtles may succumb as a result of such interference to a variety of causes. Their wrong way trek may cross busy roadways and parking lots, or lead into dense grasses and other underbrush. Hundreds can die under the tires of vehicles or become hopelessly lost amidst vegetation, only to perish the next day from heat, desiccation, or crows. It is common to find an entire group of confused hatchlings wandering aimlessly underneath a bright mercury vapor light next to a building or in a parking lot close to the beach. Hatchling turtles which are discovered in such a confused state should be collected, placed in a bucket or box, and immediately taken to a nearby darkened section of beach and released as a group at the water's edge. Use of flashlights for observing the little turtles should be curtailed to avoid continuation of their confusion.

Owners and managers of illuminated beachfront buildings within the subregion have been continuously reminded via press or television coverage of the problems that their lighting may cause to postemergent sea turtles during their critical dash to the water. Many private homes and resorts have conscientiously dimmed, or even eliminated, unnecessary bright outdoor lighting. Coastal County and City governments in Florida are adopting lighting ordinances designed to enhance survival of hatchling sea turtles. Sea turtle protection will be discussed in Chapter 22.

17
Beyond the Nest

For decades, sea turtle researchers wrestled with the question of just where did little turtles go once they reached the water? They virtually disappeared. The early life history of the loggerhead and other sea turtles following their dangerous nighttime trek to the surf has been most appropriately named "the lost year." It was only very recently that bits and pieces of information were sorted together to help provide some of the answers to this perplexing mystery.

Hatchling loggerheads, on bursting to the surface on their natal beach, immediately begin a strange initial outburst of action which helps to move them across the beach through the often violent surf to an unknown distance away from land. This aquatic activity consists of near constant swimming, in a rather precise offshore direction, along with brief intermittent periods of rest. The hatchlings are believed to make considerable headway during the three to five day swim frenzy period. Some researchers have attempted to calculate the distance that these young turtles may move during this very reliable age-old beach escape behavior. This work was conducted under laboratory conditions and although the data may be somewhat skewed, it appears that neonate loggerheads can move up to fifty miles offshore during their early swimming endeavor. The functional body energy required for a hatchling to sustain this activity must be considerable! Observations support the consensus among investigators that hatchlings take no sustenance during this strenuous swim as this mechanism ejects them, almost forevermore, from the land. A sufficient supply of nourishment is slowly absorbed from the residual egg yolk inside the wall of the plastron.

Today, just about everyone working with sea turtles is convinced that their swim frenzy mode transports hatchlings far enough offshore where if they make the journey safely, they reach huge drift lines of surface-floating marine algaes. The predominant floating species distributed near subregional nesting beaches is Gulfweed, a group of brown marine algaes of the genus *Sargassum*. Several pelagic species of *Sargassum* occur in the Gulf of Mexico and all of these develop massive floating accumulations when the right drift and current conditions exist.

There is strong evidence to suggest that young loggerheads do become associated with the deep water *Sargassum* community. Hatchling loggerheads maintained in captivity respond positively to the introduction of Gulfweed into their tanks. During development of this turtle/Gulfweed association hypothesis, researchers observed that the behavior of neonate loggerheads was altered when *Sargassum* was introduced into their tanks near the termination of the swim frenzy behavior and they freely entered, or climbed atop, the algae mat and suspended their swimming activity.

The legendary Sargasso Sea of the Atlantic Ocean is so named because of the great monotypic grouping of Gulfweed between latitude 20 degrees and 35 degrees North, and longitude 30 degrees and 70 degrees West. In this region, enormous masses of Gulfweed have become trapped because of major oceanic currents and drifts. In all probability, young marine turtles of most, if not all, North Atlantic species spend their preadolescent years inextricably associated with this Gulfweed community.

Once hatchling turtles reach Gulfweed driftlines, they climb aboard and become pelagic drifters. Their needs are met, in terms of food and cover, by the very specialized ecological community found within the *Sargassum*. A complex array of organisms found mixed with the algae provides a variety of food materials to the liking of the tiny sea turtles. *Sargassum* provides a haven for a host of shrimp, crabs, and other forms of invertebrates, and several types of small fish which thrive amidst the floating algae. Hatchling loggerheads become dependent on the forage that is passively carried along with them for some time, perhaps a year or two or even more, until it is possible that their dietary needs are no longer satisfied in the dense Gulfweed float that has been their home.

Loggerhead turtles, believed to be in the three to four year old size class, are found in Atlantic waters around the Azores Islands. Where this population of small loggerheads originated is speculative, but they are probably juveniles which have left the Sargasso Sea, entered a secondary type of developmental habitat and are beginning another stage of their complex life history. When one considers that hatchling and juvenile loggerheads are seldom found in the littoral waters of North America, one must speculate further that they remain passively associated with the pelagic lifestyle which they found earlier in the Sargasso Sea. An exception to this is the occasional small post-hatchling which is thrown ashore because of some turbulent storm event. Loggerhead turtles begin to show up in Florida waters at the subadult size class, weighing about eighty pounds. Marine turtle researchers operating in the Indian River, on Florida's Atlantic coast, sometime work with significant numbers of this size class. My records show very limited data on turtles of this size in the subregion. A

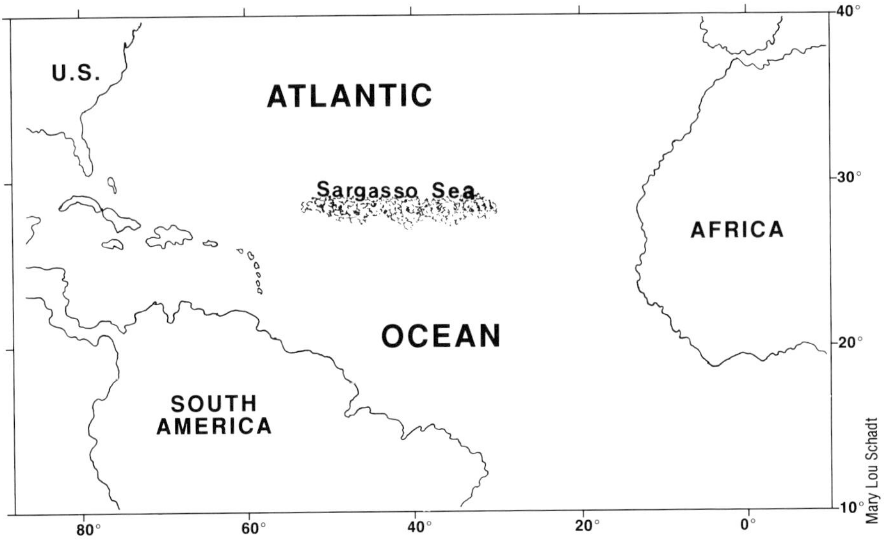

Figure 41. The Sargasso Sea.

few loggerheads in the eighty pound range occasionally are caught by commercial gillnetters, shrimp trawlers, and by sport fishermen in Southwest Florida. This size is infrequently represented in stranding statistics.

I recall one subadult loggerhead which was regularly caught from the public fishing pier at the Sanibel Lighthouse by recreational hook and line fishermen. On several occasions, I was called on to assist some astonished fisherman with the chore of removing line, bait and hook from an agitated and feisty young loggerhead. This gave me the opportunity to tag and periodically examine the individual. For months, this turtle regularly was caught at the Sanibel pier and the public fishing pier on Estero Island, some four miles away.

If hatchling loggerheads produced on the barrier island beaches of the subregion reach drifts of Gulfweed far enough offshore in time, they may well end up somewhere near the mid-Atlantic. Once they are free from any major interactions between storms, littoral currents and tides, and have reached the relative safety of the pelagic Gulfweed zone, young turtles are picked up by the peculiar offshore currents and eddies ultimately reaching the Gulf Stream. This major, fast-paced ocean current helps to propel the quick-growing turtles to their interim destination—the Sargasso Sea. Following a year or two of relatively easy living in mid-ocean, they disperse and travel on in a randomly determined route. Once away they may visit or slowly skirt the Azores or the African coast, or briefly visit other eddy-influenced ocean areas before they are committed to make perhaps a westerly and final transatlantic crossing.

It has been suggested recently that Gulfweed may not be a completely safe haven for post-hatchling sea turtles because of predacious pelagic fish that cruise beneath the matted algae. It is true that one can catch these fish by casting bait or lures beneath or around the perimeter of dense *Sargassum,* but fishermen are not as successful when their fish attractants land on the floating mats. If hatchling sea turtles are situated atop or somewhat inside the *Sargassum,* their safety is probably assured.

The time element that may be involved for a wild loggerhead to grow from hatchling size to that of an eighty pound subadult will probably remain unknown until a sufficient number of marked, known-age turtles are recovered. Presently, without a proven technique to mark and identify them at recapture, the time span required cannot be accurately determined. Captive specimens which are properly maintained with good husbandry techniques grow surprisingly fast. For example, Eve Haverfield kept a *blind* loggerhead hatchling in a child's wading pool at her residence for two years and two months. Eve and I took this remarkable specimen to the Clearwater Marine Science Center for permanent housing in December, 1986. On the day it was transferred, the turtle weighed twenty-six pounds. According to Dennis Kellenberger, Director of the Center, by December, 1988, this specimen had attained a weight of eighty-five pounds. It certainly is conceivable that healthy, wild loggerheads can grow just as fast, if not faster; however, it is the consensus among sea turtle biologists that the growth rate in the wild is much slower.

Subadult loggerheads may only be six to eight years old when they return to the coastal waters of the Americas and enter the population. The age at which *Caretta* matures in the ocean eludes us but, as I mentioned earlier, it is reasonable to assume that sexual maturity is reached in about fifteen years. The smallest egg laying loggerhead I have ever encountered on Sanibel Island measured 32.25 inches (81.9 cm), curved carapace length.

Longevity of *Caretta,* in wild populations, is unknown and few published records document captive life spans. A loggerhead lived for over thirty-three years in a European aquarium, but there is no record of its age when it entered captivity. I estimate that some of the older Sanibel Island nesting females are in the vicinity of forty years of age. This assumption is based on acceptance of a fifteen year maturation phase, followed by a calculated thirty-two year reproductive period. This reproductive life span is based on that suggested for the loggerhead turtles nesting on Little Cumberland Island, Georgia, where an estimate was developed by statistically modeling the reproductive rates of that population. I assume that the reproductive viability of subregional *Caretta* is much the same. There is probably also an age at which loggerhead females become senile, reproductively, and no longer visit land, but this too remains unknown. Beyond this level of their life, should they continue to escape life-threatening elements of their environment, a few loggerheads may reach the ripe old age of sixty or even seventy years.

18
Beach Selection

One of the more controversial issues in the biology and life history of sea turtles, is comprehension of the group's feat of nesting site fidelity. Generally, the reproductive life of sea turtles has been surrounded by romanticized, embellished tales. Traditional folklore implies that each and every one of the world's female sea turtles returns to reproduce on her natal beach because of some indelible code received during development, or some instinctive reasoning acquired as an embryo or hatchling. Long before I ever heard or read any authoritative arguments supporting this hypothesis, I was told with conviction on the part of the tellers, that subregional loggerhead turtles returned to the beach where they originated. I gleaned this tidbit of information from long-time commercial fishermen who regularly plied subregional waters and who had a fundamental knowledge of sea turtle biology.

Years later, the question of this impressed behavior was argued more scientifically, and some outstanding sea turtle people developed elaborate theoretical answers that paralleled and even gave credence to this established folklore. The earliest professional discussion about sea turtle beach imprintation that I recall in the literature, centered around undocumented reports that hatchling sea turtles were somehow psychologically branded as they raced across a beach and entered the water. This sounded good and it gave some credibility to the fishermen's notion that loggerheads returned to the same beach where they had hatched. Other factors also contributed to acceptance of this purported mechanism. Early tagging of sea turtles, especially the work on the Caribbean coast of Costa Rica by Archie Carr, produced evidence that this region's nesting population of green turtles nested only on that specific beach. So, the theory, with such splendid documentation, was further developed and accepted as a generality and became part of the unique life history of all sea turtles. If green turtles nested with such site specific tenacity then it stood to reason that this was a result of momentous fixation and that they had somehow received the exact location code early-on.

More recently, some sea turtle researchers have suggested that neonate sea turtles may be imprinted while they are developing as embryos inside the egg cavity. These workers have postulated that the chemical composition of the beach soils which surround the developing eggs is the recollection stimulus which provides for future response. Female turtles would attain maturation, travel to a specific area, mate, and search out a specific nesting beach which they identify by cueing to the chemistry of the beach. Other researchers have discarded the in-nest or beach crossing concepts and consider imprintation to be the result of a later adult recognition response of site cueing to the waters of a nearby estuary. Unforgettable chemical components, specific to the waters near their developmental beach, would be stored in an individual's memory at the instant a female hatchling entered the surf. These identifiable water quality standards are then recalled when a mature female, which has entered its individual cyclic reproductive period coincidental to being within its recall territory, finds that she is in the right place at the right time.

Recently, it has been suggested that beach selection may be a social function based on unknown turtle communication skills and even a learning experience. Under provisions of this concept first-time nesting turtles arrive randomly off a nesting beach and are recruited by experienced turtles into the nesting population. These novice turtles are supposed to follow veteran females to the beach for egg deposition, and there is some data to indicate that group nesting occurs in *Caretta*. We once watched on a Brevard County beach three loggerheads leave the water together; the first two were so close that their shells touched, and the third was trailing the pair by a few feet. The trio ventured a few yards up the beach and, for no apparent reason, all turned at the same moment and false crawled. There were no other turtles in view on that particular section of beach when this amazing observation was made. I have only observed this behavioral phenomenon once in my career with loggerhead turtles.

Another recent concept has been developed to explain why mature sea turtles do indeed return to the same beaches after absences of one, two, and three or more years. This latest attempt to resolve the question of beach selection has considerable merit, although some aspects remain unanswered. It dispels the requirement that turtles are inextricably tied to their ancestral beach because of any early chemical or psychological imprintation. It is conceivable that on entering the water the little turtles are carried randomly away to become transients in oceanic systems until they become sexually mature, mate, and have to locate a first-time beach site on which to land and accomplish egg deposition. If this first nesting venture is successful, the satisfied individual imprints to that specific beach for her future reproductive needs. A remarkable explanation, although it does not answer the question why those Costa Rican green turtles have been found nesting nowhere else. After having been tossed and turned, tagged, and measured for generations, these turtles continue to cling to that beach despite their traumatic experiences with researchers.

When applied to the loggerhead turtles of the subregion, this latest concept of the successful nesting relationship to nest site specificity justifies discussion. Interesting evidence relating to initial site selection, and later, repeat use of the general site, becomes apparent as I review tag return data. The time frame and location elements for postnesting Sanibel Island loggerhead turtles from waters of the Gulf of Mexico warrant examination. Many of the reported postnesting observations of known tagged turtles, not only Sanibel Island individuals but animals from other subregional nesting beaches as well, apparently move north from these beaches following termination of their respective nesting season. This northerly movement is sometimes discernible during typical internesting intervals when general observational location sites are reviewed. Postnesting is defined as the time following reproduction and the period of time elapsing until an individual turtle nests in an ensuing season; for example, one Sanibel Island loggerhead, CR 610, was brought aboard a shrimp boat which was trawling near the Chandeleur Islands in Louisiana waters. The capture occurred seven months after this individual had been tagged on Sanibel Island. The loggerhead was released alive after a crew member had noted the tag number, recorded the return message on the opposite side of the tag, and reported the incident to me. Fifteen months later, and twenty-two months after being tagged, this same turtle was caught and released alive by a commercial mullet fisherman near Indian Key, Collier County, Florida, some sixty-five miles south of Sanibel Island. A little over two months later CR 610 was encountered nesting again on Sanibel.

Admittedly, the quantity of this type of data is scant, but I suggest that following a successful nesting season some of the Sanibel Island *Caretta* follow the Florida coast north and ultimately, if they make a safe passage, return to subregional waters after they have circumnavigated the coastal waters of the Gulf of Mexico. Other Sanibel loggerheads have been caught incidental to fishing operations offshore of the Texas and Yucatan, Mexico, coasts. The return trip from Mexico involves deep water navigation since individuals must cross the open Gulf, perhaps in leap frog fashion—Mexico-Cuba-Florida. These data provide some degree of credibility to what, in my opinion, regularly occurs although not customarily for all individuals of the subregional population. Some subregional loggerheads move south and then well up the eastern seaboard of the continent during postnesting travels.

19
Adult Mortality

Each year, tens of thousands of marine turtles of most size classes are killed along the coast of the southeastern United States. Mortality occurs in a variety of ways. Some turtles are lost to historic natural predators and disease, while others meet their demise because of interaction with humans. The Sea Turtle Stranding and Salvage Network, administered by the National Marine Fisheries Service, collects stranding information on sea turtles which wash ashore dead and alive. Unfortunately, the greater preponderance are the former, and are then reported to the respective State Coordinators. In 1980, 1,800 dead sea turtles, mostly loggerheads, were reported, and in 1987 over 2,000 sea turtle carcasses, again predominantly *Caretta*, were reported to the Network.

Illegal harvesting of loggerhead turtles on nesting beaches still occurs on the more remote barrier islands of the subregion. Occasionally, evidence is found that indicates turtles were slaughtered on their nesting beach. Operation of motor vehicles on the barrier island beaches of Southwest Florida has been curtailed by local legislation and has reduced the opportunity for many would-be poachers to be afield. Now, the more persistent of these low-life individuals slowly operate their boats parallel to the beach, just outside of the surf line, without running lights, searching for turtle crawls or turtles on the beach. When they find a turtle and evaluate the possibility of landing without being observed, these dastards wade to the beach, dispatch the turtle, drag it to the boat where it is loaded and transported to a predetermined site. Or, the turtle may be taken offshore, butchered, and the remaining parts thrown overboard. However, dismembering a loggerhead is time consuming and the chance of being apprehended can be great if sly wits are not used. Some individuals have been doing this despicable act each summer for most of their lives and have never been caught. May their boats sink far offshore, and the only flotation device to come within their grasp be a passing loggerhead turtle, about to make a deep dive!!!

I, personally, have found the remains of recently slaughtered loggerheads on Sanibel Island, Upper Captiva, spoil islands near Matlacha Pass, in Lee County, and on Panther Key, in the Ten Thousand Islands of Collier County. The most recent evidence of a loggerhead turtle being taken from the Sanibel beach surfaced in June of 1978. While I was on turtle patrol, heading north and almost to Bowman's Beach, I noticed a flashlight shining ahead and didn't think much of this until approaching a little closer in the Jeep. I was operating without lights because a beautiful near-full moon illuminated the white beach well ahead. Suddenly, a light was extinguished and two people ran off toward the vegetation, leaving behind something large on the open beach. As the distance narrowed, I could see that the object on the beach in front of me was an upside down loggerhead turtle. The turtle was dead and the violators had already started to butcher the animal for division of the preferred body parts. Had I arrived earlier the perpetrators might have been frightened off sooner and the turtle's demise could have been prevented. I'm sure that because I have been consistently out on the beach during the nesting season working with loggerheads, my presence has been a deterrent to similar acts many times.

Not long after I moved to Sanibel, I was commissioned a Deputy Wildlife Officer by the Florida Game and Fresh Water Fish Commission. At the time this was a reciprocal type of appointment bestowed by the State to employees of the U.S. Fish and Wildlife Service who had been delegated law enforcement authority by that agency. In return, the Fish and Wildlife Service commissioned State of Florida Wildlife Officers as U.S. Deputy Game Wardens. I held this commission until the late 1970's when the program was rescinded. It was not until the loggerhead turtle joined the ranks of the Endangered Species List, in 1978, that my Federal law enforcement authority was extended to include the loggerhead turtle, its nests and eggs.

In my long career, I only apprehended one group of would-be poachers and, in retrospect, this encounter was rather comical. Well after sunset one night in June of 1960, I had left the Lighthouse and was slowly travelling west along the water's edge. My wife and daughter and a few friends were with me for a night out on the beach. In those days, before most homes on Sanibel were air conditioned, including mine, a breezy nighttime ride on the beach in a Jeep with lowered windshield was also just about the only way to escape the torment of sand flies and mosquitos on otherwise still nights. We drove along the damp beach without lights since the full moon provided ample illumination for me to negotiate the familiar route. The moon's reflection danced from the far-off horizon, shining inshore across the gentle waves which rolled silver-hued onto the beach. Visibility was excellent and any existing crawls made by turtles which had emerged the night before, or this night, would not go un-noticed.

We had not been gone long—less that two miles—when I suddenly discerned the silhouette of a familiar vehicle ahead of us with two unsteady figures trying their best to stand at the rear of their station wagon as a third person ran across the beach and disappeared into the dense upper beach vegetation. As the distance between us closed, I saw an upside down loggerhead turtle off to my left. A taut heavy line led from one of the turtle's front flippers to the bumper of the well-rusted Plymouth. I stopped just short of hitting the rope, turned on the headlights and leaped out of the Jeep. I quickly checked to see if the turtle was dead or alive, or dying, and walked over to the two very intoxicated persons who also knew me. The turtle, very much alive, resumed activity and heaved flipperfulls of sand everywhere, covering Jeep and occupants, would-be poachers, and me. The inebriated husband and wife realized they had been caught red-handed and made a few incoherent remarks to that effect. I told them they had a problem as I took my pocket knife from my jeans and proceeded to cut the rope from the turtle and the vehicle's bumper. I then rendered the remaining line into several short lengths—the latter was done with much pleasure on my part, I'm sure. I turned the turtle over and watched the lucky creature make her getaway as she disappeared into the dark waters of the Gulf of Mexico. I advised the two culprits that they should leave the beach and I would visit with them the next day.

Since the turtle was unharmed, the decision was made that there would be no legal follow-up and the individuals would not be cited or arrested for harassment or attempting to take the loggerhead. During my visit to their home the following afternoon, when I was certain that they had regained at least partial sobriety, I suggested in explicit terminology that I never wanted to encounter them on the beach again at night—and I never did. They were also advised that this kind of act was no longer going to be tolerated on Sanibel. It was to be three years until the Lee County Commission would enact legislation that prohibited operation of motor vehicles on the beaches of barrier islands in the county. Later, after its incorporation in 1974, the City of Sanibel adopted a similar regulation. Of course, bonafide research vehicles and those of law enforcement agencies are excluded from provisions of these ordinances.

There are numerous references in literature, as well as discussions and comments by self-proclaimed epicures, as to the poor palatability of loggerhead turtle flesh. When compared to the meat of the once popular green turtle, that of *Caretta* is supposed to be inferior. This may be true, but when only loggerhead flesh is available, and the consumer has nothing else to compare it to, there is no difference. Old time residents, anyway, who have actively harvested loggerheads, think their flesh is very good and have developed special recipes for its preparation. I must confess, I have once eaten loggerhead turtle; however, at the time, I did so unknowingly and did not learn the origin of the repast until later. I have since realized that this was done deceitfully.

About forty percent of the nesting loggerhead turtles encountered on

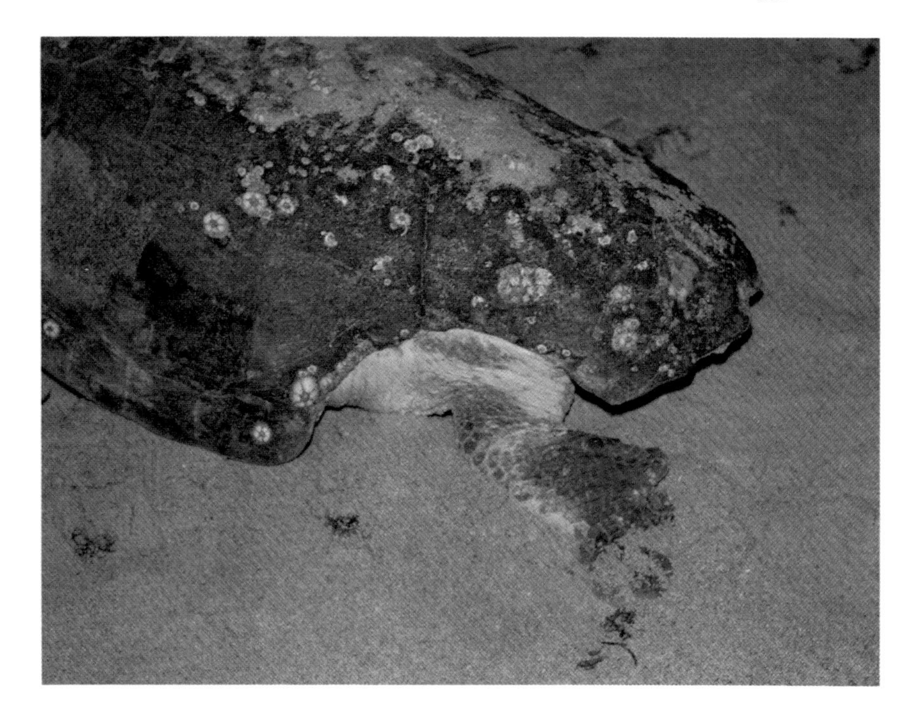

Plate 33. Results of an earlier shark attack. Such dramatic shell injuries or limb amputations are very common among loggerhead turtles.

the beaches of the subregion have been mutilated to various degrees by sharks. Their flippers, both front and rear, are frequently damaged—from crescent-shaped partial amputations of the extremities to complete removal of one or more appendages. Similar bitten areas are found along the carapace margins on some individuals. Some of these injuries appear to have been extremely severe and I have often been amazed at the hardiness of those animals that survived wounds that in my view should have resulted in their death.

Stranded turtles of both sexes frequently show signs of recent shark attack. This is evidenced by the clean crescent bites with all tissue lost, and the presence of overbite teeth indentations on bone, or torn flesh. Of course, with stranded specimens, if the injury appears to be new, it is almost impossible to ascertain if sharks are responsible for the turtle's demise, or if one of the predacious elasmobranchs simply encountered a dead turtle and seized the opportunity to consume parts of the turtle post-mortem.

Reports of shark and loggerhead encounters have been occasionally conveyed to me by observers who had the chance to see the interaction firsthand. The most recent of these occurred a few years ago and was witnessed by a sea turtle enthusiast whom I later had the pleasure to meet. While on board a party boat, which was situated for grouper fishing about fifty miles to the west of Estero Island, she and other passengers noticed the calm of the Gulf broken by an adult loggerhead turtle moving full speed at the surface and heading for the boat. As the turtle dove beneath the vessel, people lining the rail could clearly see the speeding shape of a large shark in close pursuit. When the turtle cleared the hull and surfaced, the shark closed in and began literally to tear the loggerhead apart within view of a couple of dozen awed fishermen and women. The scene soon climaxed as the turtle's headless and flipperless body settled deeper into the Gulf, and the shark too disappeared into the depths.

Some people have reported observing similar situations and have related that loggerheads will attempt to counter the aggressive advances of a shark by tightly maneuvering and maintaining the carapace exposed to the attacker in a shield-like fashion. This action certainly would provide some level of protection from such fierce predators if the turtle had the stamina to maintain its defense.

A few nesting loggerheads with severely damaged or missing rear flippers have been encountered on subregional beaches over the years. Such handicapped turtles often make several feeble attempts to excavate an egg cavity. If the palmate section of a flipper is missing, they are unable to remove soil with that particular appendage. When an excavation fails, the turtle will often move ahead a few feet and try again. Rarely when a hole cannot be dug an individual will deposit the eggs on the surface of the beach, and most of these will be ultimately broken during the concealment stage of nesting. A few times field workers within the subregion have actually dug the egg chamber for a flipper-damaged loggerhead by hand, using their palm or a section of bivalve.

Male sea turtles which range through the waters of the Western Hemisphere never touch dry ground after they escape from the sand as hatchlings. In the Pacific Ocean, a population of green turtles do visit islands to the west of Hawaii at a place known as French Frigate Shoals and move ashore to bask. I have observed only one live adult male loggerhead very close to a dry beach in the subregion and this specimen had been the victim of a shark.

Early in the morning of June 10, 1960, back in the days when I was young and foolish enough to stay out on the beach all night long, seven nights per week, all summer, I discovered an adult male loggerhead during a routine patrol who was feebly attempting to move out of the surf on the west end of Sanibel Island. This individual was bleeding profusely from shark bite wounds on both front flippers. Later that morning when I passed the site again, the turtle was nowhere to be found, and the individual never stranded on the Sanibel beach.

In the late summer of 1971, the Gulf and estuarine waters adjacent to Sanibel Island were affected by a major Red Tide bloom. A Red Tide results when a resident dinoflagellate, *Ptychodiscus brevis*, a single-celled phytoplanktonic algae, multiplies rapidly above its normal population level of about six thousand per quart of seawater to numbers that literally turn the water a rusty red and significantly alter its consistency. During the more severe Red Tide blooms, as this one was, the water's viscosity resembles skim milk. A quart of Gulf water may contain six million of the dinoflagellates.

When the cell membrane of *P. brevis* is broken, a powerful neurotoxin is released into the water column. At times of severe blooms, high concentrations of the toxin result in fish kills of staggering proportions—the 1971 bloom was such an event.

Filter feeding mollusks consume *P. brevis*, as well as other dinoflagellates, and are known to store lethal amounts of the Red Tide toxin in their organs. From late June until mid-August, 1971, eighteen dead adult loggerhead turtles stranded on the Gulf beaches of Sanibel and Captiva Islands. Since this bloom was widespread, I am certain that many other dead loggerheads washed ashore elsewhere in Southwest Florida, but their numbers are unknown. To have eighteen sea turtle stranding events in the summertime is very unusual since most strandings typically occur in the subregion in late winter or early spring.

The majority of the specimens were closely examined, and I performed necropsies on those individuals which were not in a too advanced stage of decomposition. The gastrointestinal tract of a dozen turtles contained large quantities of shell fragments of the stiff pen shell, *Atrina rigida*. These very abundant bivalves are filter feeders which consolidate the Red Tide poison at elevated levels sufficient to transfer lethal doses of the virulent neurotoxin to any of the higher animals which feed on them. I am convinced that the pen shells were indirectly responsible for the deaths of most, if not all, of the dead loggerheads found on Sanibel and Captiva Islands during the summer of 1971.

The interaction between commercial and recreational fishing activities and sea turtles often results in frustrated fisherman and major turtle mortality. Turtle catch is a byproduct of a combination of fishing technology and the demand for increased productivity from the methods used. In efforts to increase their catch shrimp boats operating in the offshore waters of the subregion have adopted gear changes to meet those demands. Some of the larger shrimp boats now drag as many as three trawl nets on each side of their vessel and have extended their drag time.

Despite claims to the contrary by the industry, shrimping is incidentally, yet directly, responsible for much of the decline in marine turtle populations. The most dramatic example of this is the sudden plummeting in numbers of the Kemp's ridley population from the Gulf of Mexico, and elsewhere, over the past several decades and, in turn, from their nesting beach in Tamaulipas, Mexico. Data on stranded turtles in the subregion, more specifically loggerhead turtles from Sanibel Island, many of which certainly resulted because of shrimping operations, are reviewed elsewhere in the text.

In an attempt to reduce the high level of incidental catch of sea turtles by the shrimping fleet operating in the United States, the National Marine Fisheries Service has developed a new piece of fishing gear. The apparatus has several names, such as Trawling Efficiency Device or Turtle Excluder Device, or TED as it is being called. I like to call it the Turtle Escape Device since, in my opinion, this terminology is more appropriate. The use of this addition to each net being trawled has demonstrated an increase in shrimp catch, a decrease in bycatch which includes sea turtles, and improvement of water flow through the trawl which results in drag reduction and probable savings in fuel costs. Studies have shown that utilization of TED's have improved shrimp catch by as much as 7.5 percent, and reduced the capture of sea turtles by 97 percent.

Turtle losses because of shrimping operations are no small problem. Data indicate that 47,000 sea turtles are caught annually in shrimp trawls. Of these, 11,179 (23.8%) die in the nets from asphyxiation or drowning. Another interesting study conducted by the National Marine Fisheries Service, relative to the turtle excluding benefits of TED's, clearly demonstrated the device's efficiency. In an on-going research project, a shrimp boat has been outfitted with a pair of nets. One of these trawls has a TED installed and the other does not. After 2,800 hours of trawling time, there were no sea turtles caught in the net containing the TED, but in the net without the TED twenty-four sea turtles were caught!

Several types of TEDs have been certified by the National Marine Fisheries Service for use by the shrimp industry. Some are welded metal devices and some of the more recent models are called soft TEDs. Their role is very effective, in terms of sea turtle conservation, since they all have the ability to eliminate or help prevent large-sized, unwanted, bycatch from reaching the bag of the net. Sponges, cannonball jellyfish or turtles can be ejected by differing space configurations between the funnel bars or webbing of the TED, depending on the model being used. These bars or rope webbing are angled and deflect large, undesirable organisms through a hinged, weight-activated, trap door on the top of the TED, while shrimp pass through the deflecting partition and end up in the cod end of the trawl.

One of the approved soft TEDs has the release opening located on the bottom of the netting.

In many parts of the southeast shrimpers have not accepted the TED. This has primarily been an economic response since the TED increases the cost of outfitting a trawl by up to three hundred dollars per net. Some refuse to learn techniques necessary to properly use the devices, which would help them maintain shrimp catch levels. Improved fuel costs and the reduced on-deck sorting time required because of reduction of bycatch, will help to encourage the acceptance of the gear. Some States are providing financial aid to shrimpers for TED acquisition.

The utilization of the TED in any of its approved variations has become an important part of shrimping operations. Beginning May 1, 1989, shrimpers operating on the offshore waters of the Gulf of Mexico and southern North Atlantic, from Texas to North Carolina, with boats over twenty-five feet in length were to have installed TEDs in their trawls. Vessels under twenty-five feet are not required to be outfitted with TEDs, but must restrict their towing time to ninety minutes or less. Beginning a year later, on May 1, 1990, trawlers fishing inshore also will be required to either use TEDs or restrict their tow time to ninety minutes or less.

In August, 1989, the Secretary of Commerce did an about-face, bending to the vocal demands and demonstrations of the shrimping industry, and suspended the rule requiring the use of TEDs in waters regulated by the United States. Instead of requiring total use of TEDs, the rule was reworked to allow boats to drag for one hundred five minutes. Nets were to be aboard the vessel during scheduled one half-hour time periods. For shrimpers intelligent enough to utilize TEDs, the half-hour no fishing period was not applicable. The Department of Commerce insisted that their tow time regulations were enforceable and met the mandates of the Endangered Species Act. It seemed to me at the time, they had developed the special tow time rule in capitulation, accommodating people who are defiant and simply don't want to obey the laws of the land.

On September 6, 1989, the Commerce Department was forced to flip-flop again by a court order. This order directed the Department to protect sea turtles, and on September 8, the TED rule was back in place.

Loggerhead turtles regularly frequent "live bottoms" where concentrated shrimping operations occur. These areas generally harbor a wide range of organisms which have selected this type of niche for a variety of reasons. Loggerheads choose such bottoms because their primary food sources occur in this habitat. One such site, known among shrimpers as the "Mudhole," is located to the south-southwest of Sanibel Island. Here in this optimum habitat bottom feeding or sleeping turtles may be engulfed by the trawl nets, frantically try to escape, and if a TED is not installed, be hauled aboard the shrimp boat. If a turtle were picked up early in a drag, before the implementation of the TED or tow time regulations, it could have been trapped inside the net for up to three hours or more before being brought to the surface. As they futilely try to escape from the submerged net gear, their metabolism increases, stressing their respiratory and circulatory systems. Many of them, when harshly dumped upon the deck, are comatose or extremely lethargic because of deprivation of oxygen. Submerged turtles who have their freedom to obtain atmospheric oxygen prevented seal their respiratory system by closure of the glottis. This action, if submergence is prolonged, leads to asphyxiation and not drowning. The one hour and forty-five minute tow time, that for a short period of time was in effect, is considered by most sea turtle conservationists as excessive and too far above the mortality threshold.

Various techniques can be utilized to resuscitate seemingly lifeless sea turtles. Specimens sometimes can be revived by the following procedure: The turtle should be placed in a shaded area on the vessel's deck and situated upright, with the plastron down. The posterior of the shell should be propped up and well elevated—this helps to drain any water from the mouth, esophagus, or lungs. Occasionally, the turtle should be wet down and observed for any indications of movement. Specimens which are held for over twenty-four hours in this fashion and which show no evidence of life can be presumed dead and must be dropped over the side of the vessel. Some people report that if a short length of garden hose is inserted into the trachea and air is blown into this tube, many turtles which are comatose and appear dead can be more quickly restored. Turtles that are energetic when first decked and those that are revived must be released when the trawling gear is aboard and the vessel is in neutral. To hold live sea turtles aboard a vessel or bring them to a dock or ashore is in violation of Federal regulations and State statutes.

Longline fishing is another commercial fishing technique that is becoming more common in the subregion and is beginning to impact sea turtles. This deep water method of fishing for grouper and snapper consists of literally miles of cable and heavy monofilament line with baited hooks. Turtles become hooked after taking the bait and die on the bottom, or may become entangled in the line. In 1988, one loggerhead stranded on Sanibel that I am certain succumbed as a result of this kind of fishing. The specimen's neck and one foreflipper were severely entangled in a heavy monofilament line several yards in length.

Also that same year, a longline grouper fisherman from Marco Island telephoned and told me that he had caught one of my tagged loggerheads on his line when it was set west of Marco in about three hundred feet of water.

Another bit of frightening fishing gear that is gaining popularity is the drift gill net. It consists of a section of netting that may be miles long; it simply drifts below the surface and catches nearly everything that cannot pass through the mesh. I do not have any personal knowledge of turtle mortalities resulting from this method in the subregion, but I'm sure that it's just a matter of time. In 1989, an abandoned drift net, which held the entangled carcasses of ten green turtles and one loggerhead, was discovered in the Atlantic off the Florida coast.

Now that marine turtles are listed as endangered or threatened species and enforcement has improved over the past decade, it has created a hardship for those of us who are collecting life history information on the various species. Until the early 1970's I was kept current on incidental catch of sea turtles—greens, loggerheads, and Kemp's ridleys—in much of the subregion, by commercial net fishermen who were members of the Organized Fishermen of Florida. Today, I no longer benefit from their knowledge and experiences and these data are lost. Because of fear of penalties, the majority of the serious, hardworking, cooperative gillnetters, some of whom are my friends, will no longer share sea turtle catch information with me. This attitude is well justified from their standpoint and even though information is lost, I really wouldn't want to see the situation any other way.

At any time of the year, tens of thousands of submerged crab traps—offshore, for both blue and stone crabs, and inland waters, for blue crabs—crowd the waters of the subregion. From October until May, there are countless thousands of styrofoam trap floats, tethered to stone crab traps, dotting the surface of the Gulf of Mexico just offshore, and for miles out from the beaches of the barrier island chain. Year around, the inland waters are similarly decorated by the round floats attached to blue crab traps. Loggerhead turtles will attempt to take crabs that are entrapped, or they may be attracted by the cut bait inside the traps, and in doing so occasionally damage or destroy the trap devices. Most crabbers, like shrimpers and lobstermen in Florida, do not have kind words for loggerheads.

On April 14, 1969, I received a telephone call from a concerned Sanibel resident who informed me of a distressed sea turtle in the Gulf of Mexico about one hundred yards offshore and just south of Bowman's Beach. I loaded a small aluminum skiff into a Jeep and went to investigate. When I paddled out to a crab trap float making headway on top of the water, a very large loggerhead surfaced next to the boat. I managed to grab the line and tugged and paddled until, exhausted, I was able to reach the beach. Once there, several people aided me in pulling out the thirty pound stone crab trap, with an attached male loggerhead turtle! Upon examination, I found that the polypropylene line between the float and the trap had encircled the turtle's neck several times. I cut the rope free and discovered a deep, raw, ulcerated groove where the ever-tightening line had been, apparently for some weeks. With help, I loaded the turtle into my vehicle and kept it overnight underneath my residence at the Sanibel Lighthouse. The next morning, the turtle appeared to be alright, so I measured, tagged and released the specimen at the Lighthouse Gulf beach. Incidentally, this was the first, and only, adult male loggerhead that I have ever tagged.

Since this incident I have found and been brought several sea turtles that were wrapped up in crab lines. Some of these were dead, stranded individuals, while others were weak and near-fatalities. Crabbing appears to be a serious threat to survival of the loggerhead and other sea turtles.

An increasing number of turtle carcasses that strand on the beaches of the subregion have been struck by boats. The telltale deep concentric propeller gashes across a turtle's head or carapace may lead one to suspect the cause of death; however, unless the specimen is very fresh, it is difficult to judge whether the impact occurred post-mortem, or was the sole contributing factor which led to the turtle's death. As high speed pleasure boating traffic continues to escalate, the frequency of related sea turtle fatalities will also.

It is unlikely that idle or slow speed zones for boats, similar to those which have been established for the Florida population of the West Indian manatee, *Trichechus manatus latirostris*, will ever be created for marine turtles in the subregion. Any future critical habitat for sea turtles which may be delineated would be restricted to nesting beaches. Powerboat races and ultra high speed regattas are gaining in popularity. One of these is the Lee County Offshore Regatta which has been held in late spring in the Gulf waters off Sanibel Island. Coincidental to the very heavy and fast boat traffic, a dead boat-maimed loggerhead or two has floated to rest on area beaches almost every year the Regatta has been held, and always within a few days following the races.

In some coastal areas in the southeastern United States, the heads of hydraulic dredges occasionally may kill sea turtles of various species and sizes. Shipping channels leading to ports in Tampa Bay are continually being dredged, while other dredging in the subregion is done on shorter channels and on a less frequent maintenance schedule. Loggerheads and other species of sea turtles are probably sucked up into these devices, but I have no data to indicate how serious this problem is to subregional sea turtle stocks.

Despite international efforts to curtail the discharge of petroleum products into the world's oceans and estuaries, such pollution still occurs. Oil spills adversely impact sea turtles and are a significant factor in the mortality of subadults. Occasionally, small individuals, after having somehow contacted considerable amounts of petroleum solids, are found cast up on beaches along the eastern seaboard with their mouths and nostrils sealed shut by tar-like compounds. This probably happens when the turtles actually ingest floating or sunken tar balls, or when they surface for air underneath floating heavy petroleum such as Bunker C. Rehabilitation of these turtles can be accomplished if they are found in time.

There are no available records to document this problem within the subregion, and the last oil spill which caused moderate environmental damage occurred in Tampa Bay in 1978. This spill, of about 40,000 gallons of Bunker C and Diesel fuel, did not escape into the Gulf of Mexico and, other than affecting a few birds, resulted in no apparent long-term damage.

Plastic products which enter estuarine and oceanic ecosystems as litter often negatively impact marine turtles. Once floating plastic bags will invert and bob along at the water surface, foraging sea turtles may mistake these clear containers, or sheet goods, for jellyfish and consume them. The plastic later forms an obstruction that seals the animal's gastrointestinal tract and the individual will slowly starve to death. I implore all boaters and fishermen to discard their plastic bags and other items of litter in an appropriate receptacle on shore.

This problem occurs infrequently in loggerhead turtles, but is a common occurrence in leatherbacks. Both species prey on jellyfish—the latter much more regularly than the former. Fifty percent of the stranded leatherback turtles which come to rest on the coast of the northeastern United States contain plastic in their viscera. Some that have been examined were reported to contain enormous sheets of polyethylene film. It is the opinion of researchers who performed necropsies on these leatherbacks that the presence of plastic materials was responsible for their deaths. On necropsy a few stranded loggerheads from the Sanibel and Captiva area were found to be blocked internally by plastic fragments.

Florida sea turtles, inhabiting sounds, bays or other inland waters are subject to immobilization and mortality because of hypothermia as a result of low water temperatures. Some species which range north along the Atlantic seaboard to the Cape Cod region, are also severely stressed during winter months when they occupy similar shallow habitat. Lowered water temperatures result in lethargy and death if a turtle's exposure is prolonged and its core body temperature reaches the vicinity of ten degrees C.

In June, 1976, I conducted a short-term study relative to the body temperatures of a small series of randomly selected postnesting loggerhead turtles. During the month, fifteen animals were intercepted and restrained after they had nested and were enroute to the water. A flexible probe thermometer was inserted twelve and one half centimeters into the cloaca of each female and the temperature recorded. Coincidental to taking body temperatures, air and surf temperatures were also noted. Air temperatures ranged between 19.5 and 25.0 degrees C, with an average temperatures of 22.0 degrees. The surf temperature varied from 24.5 to 27.0 degrees, and averaged 26.3 degrees C. Cloacal temperatures ranged between 27.5 and 30.5 degrees. The mean body temperature of the loggerheads sampled was 28.5 degrees C.

The environmental factor most closely associated with the body temperatures of the sampled specimens was the surf temperature. These animals had been out of the water for a minimum of an hour and may have cooled slightly, but still were over two degrees warmer than the surf, and averaged six and one half degrees higher in temperature than the air.

During extremely cold winters, green turtles and, to a lesser extent loggerheads, are severely impacted in the Indian River ecosystem on the Florida east coast. Great efforts are made by conservation-oriented individuals, groups and agencies, to collect stressed turtles at such times. The animals are placed in warmer pools and tanks until hypothermia-produced symptoms are alleviated and natural system water temperatures return to a level which eliminates the threat of mortality. Similar weather-related events occur within the subregion. Hypothermia-stunned green turtles have been collected from Matlacha Pass and Lemon Bay in recent years. Cold-stressed turtles in adjacent waters probably went undetected or were not reported.

Hibernation, a function that overrides and eliminates hypothermic stresses, is a process that until quite recently was unknown for sea turtles. A few years ago, considerable excitement was generated among sea turtle people when an over-wintering population of green turtles was discovered apparently hibernating on the bottom of a section of the Gulf of California.

A similar condition has since been discovered among some of the loggerhead turtles which occupy the Port Canaveral shipping channel near the Kennedy Space Center on the Florida east coast. Lethargic, mud-stained loggerheads have been trawled up from the bottom of the channel during periods of extremely cold water. Some biologists are of the opinion that the stains are the result of the turtles being imbedded in the anaerobic bottom ooze, and that as cold water drastically reduces their mobility and forward movement, these animals fall, unintentionally, to rest head end first into the very soft bottom. When water temperature rises above the immobility threshold, the turtles again become active and resume their normal activity. One could conclude that these cold-stunned turtles fall, involuntarily, through the water column, sink into the bottom and do not actively bury themselves.

Recently, however, an interesting phenomena was brought to my attention that sheds some light on the ability of *Caretta* to engage in bottom burying behavior. In Florida Bay, the body of water situated between Cape Sable and the Florida Keys and located just beyond the southern boundary of the subregion, much of the bottom is densely vegetated with turtle grass. Throughout the Bay, small sandy, barren openings devoid of vegetation and known locally as "blowouts," are interspersed in the luxuriant *Thallasia-*

covered bottom. Kimberly Holmquist, a Research Assistant affiliated with the Marine Science and Conservation Center on Long Key, in Monroe County, provided me with details on a rather unique set of observations she made.

In April of 1989, while Kim was slowly being towed behind a boat over Arsnicker Banks near Arsnicker Key, to survey the bottom for marine algae, she approached a blowout just as a well-concealed adult loggerhead launched itself from beneath the surface of the sandy clearing. The turtle moved from its camouflaged resting place so quickly that Kim was momentarily frightened by the sudden violent explosion of sand from the bottom. Since that first alarming experience, she now conducts her snorkel-assisted surveys with a watchful eye and closely scrutinizes each blowout as she approaches it. There have been numerous occasions when loggerheads were seen scurrying away from blowouts as Kim approached them. In June 1989, at a water depth of eight or nine feet, she watched a small loggerhead move to a blowout, glide in and settle on the bottom. The blowout was approximately ten by fourteen feet in size. After stopping, the turtle began to make sweeping movements with its foreflippers—not unlike those made by a nesting female during nest site concealment—creating considerable turbidity in the water. As the water cleared, Kim could plainly see that the animal's head and most of its shell were completely buried—only the top of the carapace was visible.

Kimberly checked the site twenty-four hours later and the loggerhead was still buried in the blowout. She rechecked the site forty-eight hours later. The turtle was gone, but a scalloped area where the turtle had buried itself remained. These remarkable observations confirm that *Caretta* can, and do, bury and conceal themselves in the bottom, in some habitats. This raises questions on the origins of the common blowouts: Are they created and then utilized regularly by individual territorial turtles? These valuable field observations should be followed up to determine and document just how widespread this burying behavior is in Florida Bay and what size classes of *Caretta* take advantage of this habitat.

If it is eventually found that loggerheads regularly bury into soft bottoms in some habitats, with the upper part of the carapace protruding above the bottom, this behavior may explain the level of profusion of encrusting benthics residing on their shells. A sedentary lifestyle where a loggerhead turtle is partially buried in the bottom while sleeping or seeking concealment, would regulate the extent of an individual's carapacial epibiont infestation. Only that section of carapace protruding above the bottom sediment would be subject to fouling by benthics. Further, it may explain why the occasional female loggerhead is encountered which is entirely free of fouling organisms, other than algae. If an individual loggerhead turtle spends considerable time with its shell completely buried, then it would seem logical that benthics could not colonize the carapace.

Part 3

20
The Sea Turtle Stranding and Salvage Network

The Sea Turtle Stranding and Salvage Network (STSSN) is administered by the Southeast Fisheries Center of the National Marine Fisheries Service (NMFS), in Miami, Florida. I have been involved in the STSSN since it was created and endorse the program for both its data collection benefits, and as an essential tool in developing strategies which will ultimately enhance the survival of the Nation's marine turtle resource. A host of other dedicated participants throughout the coastal areas of the Southeastern United States have helped to make this Network very effective.

Stranded sea turtle carcasses are examined and limited data are collected by the authorized on-scene person. This information is then reported to a State Coordinator who, in turn, forwards the field data to the NMFS STSSN Coordinator in Miami. Reports of dead turtles originate from a variety of sources. The efforts of private citizens, governmental employees, and law enforcement agencies are responsible for formation of the STSSN data base. On Sanibel Island, when a turtle carcass is inspected, its carapace length and width are measured and the remains are closely examined for possible tags and evidence of external damage. Abraded sections of carapacial scutes or head skin and broken marginal scutes may indicate damage that resulted when a specimen was dropped from a shrimp trawl onto the vessel's deck. Deep concentric gashes atop the shell may reveal that the animal was struck by the propeller of a boat. The general physical condition is evaluated and old scars or amputations are noted. After all the information has been placed on the data sheet, the head of the individual is sprayed with paint and the initials of the observer are also painted on both sides of the carapace. The marking indicates that the specific turtle has already been examined and its removal from the beach is authorized. The Public Works Department of the City of Sanibel is permitted to handle the carcass for removal. A City crew picks up the animal and disposes of it at a prearranged site. This system has worked well over the years and the City employees who handle this nauseating task deserve a lot of credit.

The STSSN was established formally in 1980. I accepted the responsibility of serving as the Sanibel and Captiva Islands cooperator and have personally responded to the great majority of dead sea turtles that have washed ashore on island beaches, or on the nearby adjacent mainland. When I was unavailable, an associate handled the examination of stranded turtles and data collection for me. Prior to the establishment of the STSSN, I maintained, as part of my loggerhead turtle studies, accurate records of sea turtle stranding events for Sanibel Island. My records begin in January, 1959.

Many factors contribute to the number of sea turtles that strand in the subregion or along coastal areas elsewhere. An accurate number of individual turtles which meet their fate directly as a result of traumatic causes is unknown. In any one year, more specimens may be killed because of crushing boat impacts, fishing activities, or some other man-related cause than in another. Mortalities occur because of a variety of other factors too, reasons that are not controlled by man's influences or direct action. Diseases, parasites, and natural predation are factors which also are responsible for sea turtle strandings. Attrition in the population because of natural causes results in potential stranding candidates too. In the subregion, strandings are usually seasonal, with the majority of specimens stranding in late winter or early spring. See Figures 42 through 45 for a monthly distribution of individual loggerhead turtle strandings for Sanibel Island in the selected years of 1972, 1977, 1982, and 1987.

It is uncanny, but so far, loggerheads tagged on Sanibel and Captiva Islands have never been discovered as stranded animals on the beaches of the Islands. This suggests that nesting females which land on subregional beaches and are tagged, may be transients who visit local waters temporarily for the specific needs associated with procreation. These mature turtles are not present locally during the late winter and early spring stranding period, or locally tagged specimens would be represented among those stranding. Females, which are included in the subregional nesting population, probably enter waters in the vicinity of the barrier islands after the end of the usual stranding period, arriving in late April or early May, remain a few weeks and then, as discussed earlier, move on.

Late in the afternoon on June 22, 1985, I received an unusual telephone call. The young man calling who refused to give his name, informed me that he was a crew member aboard a shrimp boat operating out of Fort Myers Beach. On the night of June 10, while trawling for shrimp twenty-five miles west of Fort Myers Beach, they dislodged a dead loggerhead from their net, and removed one of my tags, CR 5098, from its flipper. The turtle was also identified by a carapace tag, but this had not been detected by the crew.

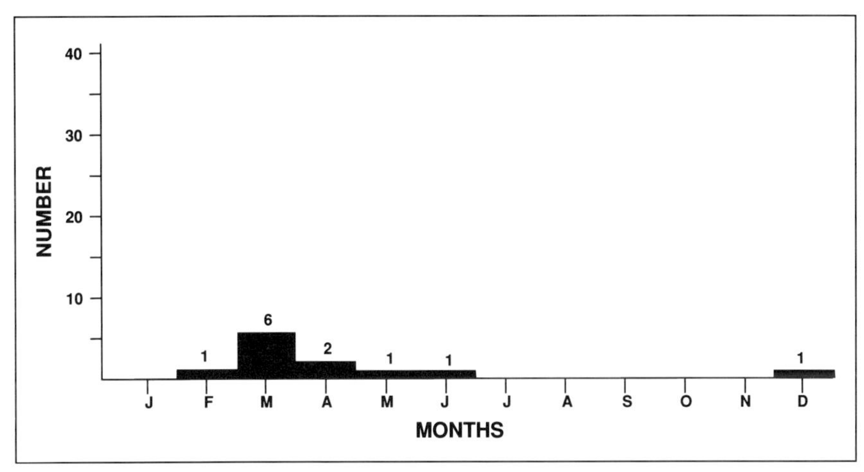

Figure 42. Monthly distribution of loggerhead turtle stranding events for Sanibel Island—1972.

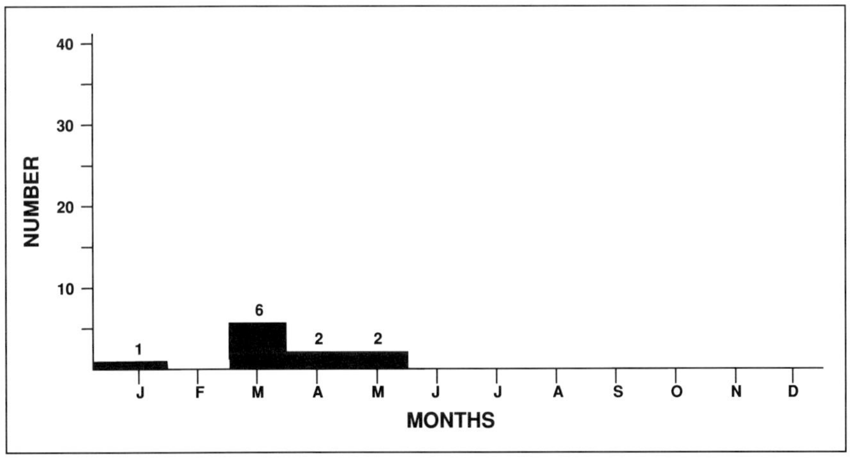

Figure 43. Monthly distribution of loggerhead turtle stranding events for Sanibel Island—1977.

The turtle was dumped over the side, and if it ever stranded it was not identified as one of mine, and I never received further word on it. I had tagged the turtle, CR 5098, SAN 122, on Sanibel Island on June 8, 1985, just two nights before it met its demise.

Overall, numbers represented in stranding reports are considered by the officials of the STSSN to be minimum stranding figures, since these only represent reported strandings, not all stranding events. Every mile of oceanic and estuarine coast in the Southeastern United States which is part of the

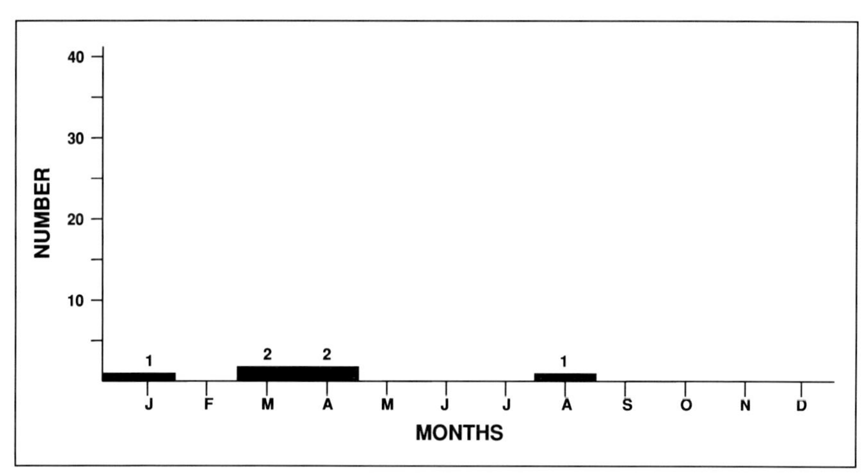

Figure 44. Monthly distribution of loggerhead turtle stranding events for Sanibel Island—1982.

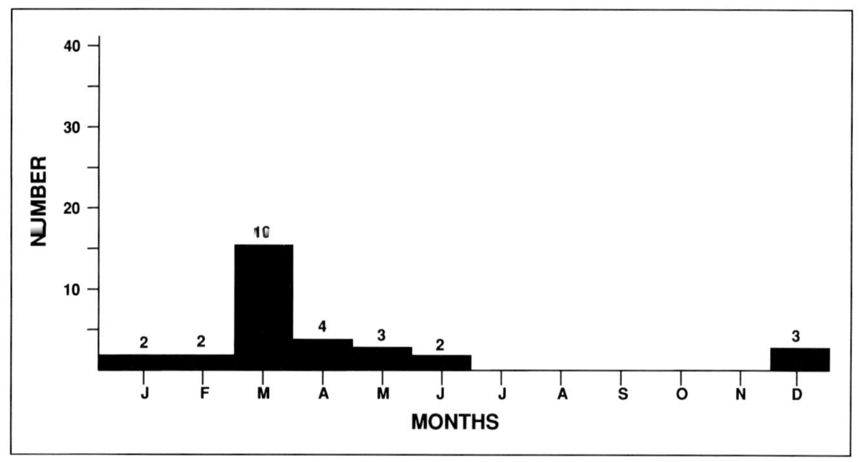

Figure 45. Monthly distribution of loggerhead turtle stranding events for Sanibel Island—1987.

various statistical zones included in the Network is not monitored for stranded sea turtles, and many animals which wash ashore may not be reported.

For purposes of demonstration of mortality levels, I have selected the years 1982 and 1987, from the STSSN data base to discuss sea turtle strandings in the subregion. For a comparative relationship, the total number of *Caretta* strandings within the selected subregional boundary, west of 81 degrees, 00′ West and east of 86 degrees, 00′ West, for these years is 47 and 153, respectively.

	COLUMN 1 Eastern Gulf of Mexico Strandings	COLUMN 2 Entire STSSN Strandings
Caretta caretta	565	1,752
Chelonia mydas	114	157
Dermochelys coriacea	5	135
Eretmochelys imbricata	9	29
Lepidochelys kempi	31	217
Unidentified	16	98
Totals	**740**	**2,388**

Table 5. Total sea turtle strandings reported to the STSSN in 1987. Column 1 reflects strandings from the subregion and Gulf waters to the north as well, while column 2 indicates total strandings for all areas of the Network.

When examining the total strandings reported to the STSSN for the years 1982 and 1987, from all the subregional statistical zones, one through eight, the results are dramatic. See Figure 46 for the locations of these zones. Strandings for these years are shown in Figures 47 and 48. The unidentified classification indicates that an examined carcass could not be positively identified as to species by the field observer.

Weather events and water temperatures can influence carcass stranding rates. Offshore winds keep carcasses away from land and, when warm water temperatures are concurrent, the decomposition rate is increased and specimens may never reach the coast. They simply sink offshore, are never observed by a STSSN cooperator, and do not become part of the data base. Cooler water temperatures tend to preserve carcasses longer. When strong onshore winds occur coincidental to low water temperatures, dead sea turtles arrive more quickly onto the barrier islands and in a less advanced stage of decomposition.

The year 1987 was a tragic year for subregional sea turtles; however, 1989 was even worse! In 1987, thirty-two loggerheads stranded on Sanibel Island alone. It is unclear if this phenomenon resulted because of a true increase in turtle mortalities or simply because of unusual climatological elements. When most of the strandings occurred in March, unusual weather patterns existed which may have contributed to the marked increase in strandings. In the early part of the month, gale force winds out of the north-northwest prevailed for several days. Turtle carcasses which were floating offshore, or submerged closer to shore, quickly appeared on the beach. A week or so later, very strong near-gale force winds out of the south swept along the Southwest Florida coast. Under this set of circumstances,

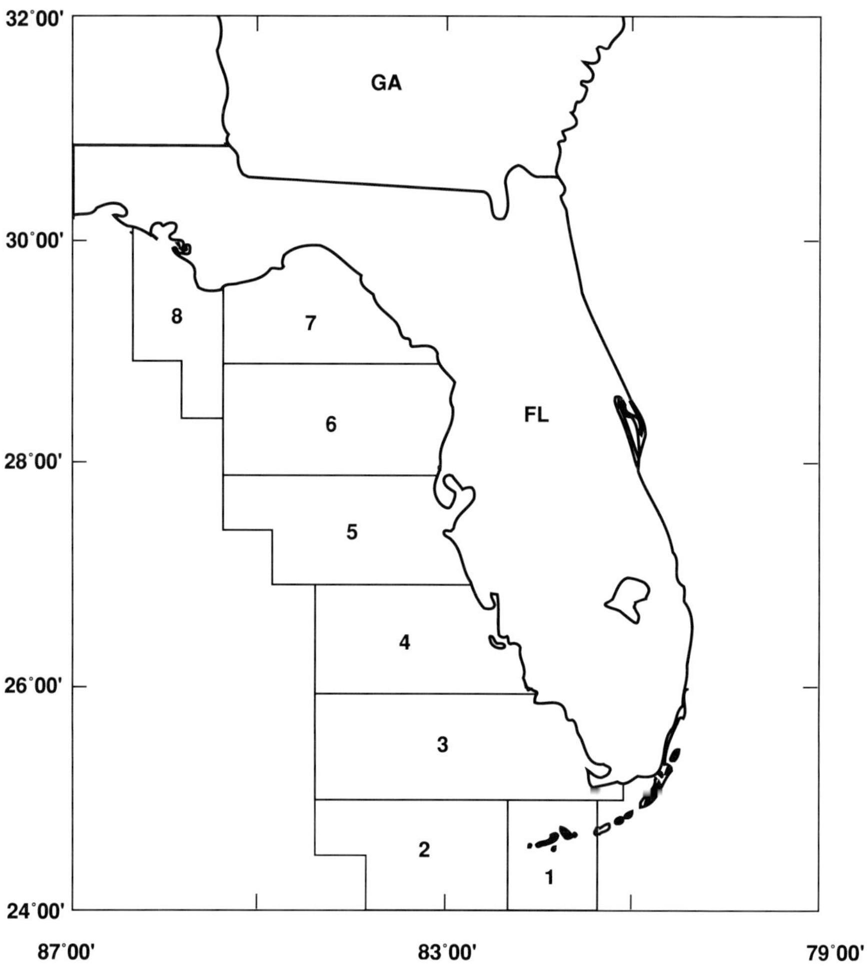

Figure 46. The statistical zones of the Sea Turtle Stranding and Salvage Network in the subregion.

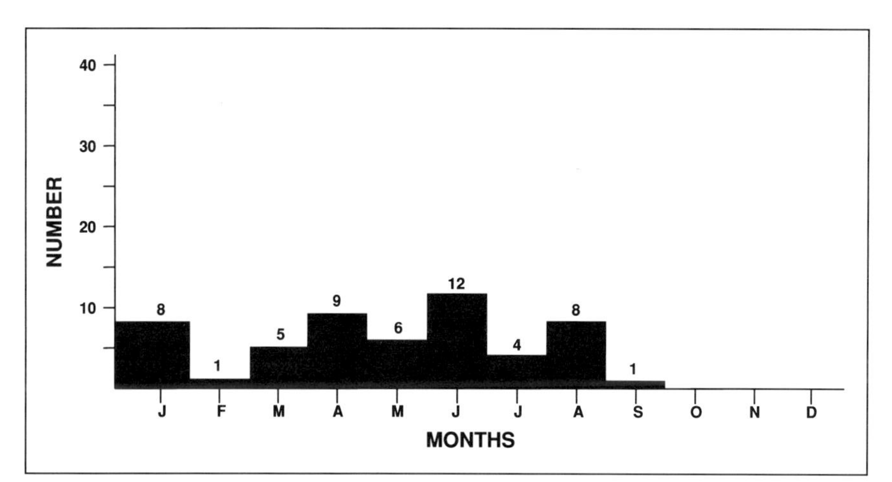

Figure 47. Monthly distribution of loggerhead turtle stranding events in statistical zones 1-8, 1982. Includes Sanibel Island data.

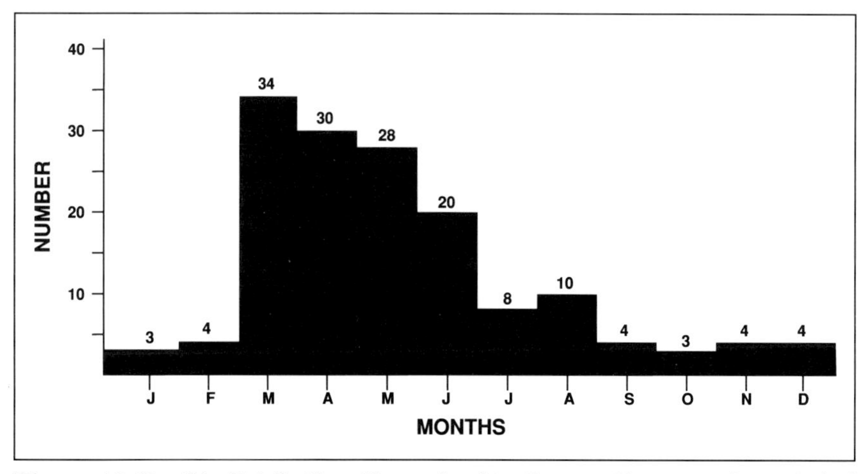

Figure 48. Monthly distribution of loggerhead turtle stranding events in statistical zones 1-8, 1987. Includes Sanibel Island data.

the geographical configuration of Sanibel Island which projects a considerable distance further offshore than the mainland beaches to the south, served as an enormous "net" and dead buoyant turtles which were being swept along by the high winds and seas were pummelled against the southern Sanibel shoreline. It is important I mention that just prior to the heavy weather and rise in strandings, there was a dramatic increase in shrimp trawling extremely close to the Sanibel beach. Some boats were observed dragging their nets within two miles of the beach. This was not customary shrimping practice, at least not by boats which normally use subregional waters. I cannot recall ever observing as many shrimp boats actively shrimping so close to the Island. While climatological conditions may have washed more carcasses ashore than normally would have stranded, the dense shrimp trawling activity so close to shore was the major factor, in terms of the total increase in the number of turtle deaths. Because these shrimp vessels were trawling so close to the shore, they were certainly responsible for the unusual number of fresh dead turtles which I examined that spring. My records indicate that in all the years prior to 1987, where I had the opportunity to monitor stranded turtles, there were only four specimens that I can say with assurance were in fresh condition. These turtles had been dead less than twenty-four hours and were still in a state of rigor mortis. Of all the other turtles I have examined, some were in the early stages of decomposition, but the great majority were extremely deteriorated. Both conditions are typical for most strandings which are reported to the STSSN.

In 1989, losses were devastating. Forty-nine loggerheads stranded on Sanibel-Captiva, and in my opinion shrimping was the overriding cause of this increase in mortality. Shrimping activity was so close in, it had become preposterous! In early April, 1989, just past 1330 hours, I personally observed three shrimp boats dragging less than one mile from the beach. At that moment, it occurred to me that perhaps it was being done maliciously to further antagonize the concerned officials and citizenry of Sanibel Island. Not only do the personalities aboard such vessels care less for the survival risk of sea turtles, they also seem to have the audacity to flaunt their contempt.

At the time, there was no demarcation line established that ruled just how close shrimp vessels could drag along the shore of the Islands. The City of Sanibel responded to the dramatic increase in sea turtle mortalities and other impacts to the near-shore bottom ecosystem. The City Council requested the Florida Marine Fisheries Commission to establish a seven mile no shrimping zone in the Gulf waters off Sanibel. On May 11, 1989, the Florida Marine Fisheries Commission instructed their staff to prepare a rule which established the line at five miles off of Sanibel, Captiva, and Cayo Costa. The line merges with an existing demarcation line, to the south of Sanibel, which extends from Wiggin's Pass to the Tortugas area. The rule became effective on October 1, 1989, and established a separation line between stone crabbers and shrimpers. This rule shall be in effect from October 1 until May 31 of each year, or the critical period when the majority of sea turtle strandings occur.

I have not included strandings of other marine turtle species on Sanibel Island in this chapter. Information on strandings of other species is con-

YEAR	STRANDINGS	YEAR	STRANDINGS	YEAR	STRANDINGS
1959	12	1970	10	1981	7
1960	0	1971	28	1982	6
1961	0	1972	12	1983	8
1962	3	1973	8	1984	11
1963	0	1974	10	1985	17
1964	0	1975	8	1986	15
1965	4	1976	12	1987	32
1966	0	1977	11	1988	12
1967	1	1978	8	1989	46
1968	7	1979	10	**TOTAL**	**317**
1969	9	1980	10		

Table 6. Total strandings of *Caretta caretta,* 1959-1989, from Sanibel Island, Florida.

tained in the appended individual accounts of those species. Listed above, in Table 6, by year, are the annual stranding totals for Sanibel Island. Loggerheads stranded nearby on Captiva Island and the adjacent mainland are not included.

The relatively low number of stranded loggerheads from Sanibel Island in the 1960's is related to several factors. In those days, when stranded turtles reached the Sanibel beach, few residents and far fewer tourists paid much attention to dead turtles. Many carcasses probably disintegrated on the beach or were buried in front of the few homes which were still occupied in the late spring. There was no public concern for sea turtles in those days, as there is today. Until 1963, because Blind Pass was located well east of its latest opening (1989), I was unable to reach without wading through difficult, or even dangerous currents, the area known as Bowman's Beach. Any stranded turtles went unnoticed on this three mile long remote beach. In 1963, Blind Pass closed and summertime Jeep patrols became possible until the opening was re-established near where it is now by Hurricane Agnes in 1972. I never made Jeep patrols specifically to locate stranded turtles, but depended on reports from others. On extremely low tides, from 1963 until 1972, I could leave the Lighthouse and drive the Gulf beach to Captiva, cross the Blind Pass Bridge, drive the roads back to my home at the Lighthouse, grab a cup of coffee, and head out again. If beach and tidal conditions permitted, I sometimes made this circle three times per night.

There are many different reactions manifested by people who encounter stranded turtles on a beach. Some are disgusted by the odor, demand immediate removal by officials, and care little about circumstances which

Plate 34. A stranded adult male loggerhead on the Captiva Island beach during the 1971 Red Tide bloom.

may have contributed to the animal's demise; however, others are very interested, concerned, and saddened.

Before a dead turtle has been left high and dry by a receding tide, water, flowing around the carcass, will move the flippers and head in a lifelike manner. When reports of stranded turtles are received, it is common for the caller to insist that the animal is alive—because it moves! There have been a few times, responding to a stranding report, I found persons sitting in the water, holding a turtle's head elevated between their legs to prevent the animal from drowning. These kind souls are astonished when I thank them for their efforts and inform them that the turtle has been dead for several days.

With local community involvement in sea turtle conservation generated with the founding of Caretta Research, Inc. in 1968, I established a local reporting network that has become an important aspect of the subregional STSSN effort. Now, a turtle which washes ashore is reported to me almost immediately and I'm sure that I have been made aware of every stranded sea turtle which has come to rest on the crescent beach of Sanibel Island since 1968.

Eighteen of the strandings in 1971 were summer events that were related to a massive Red Tide outbreak that summer, and which were discussed earlier in the text. In addition, ten loggerheads stranded at the more typical times that year.

In late 1987, with the approach of the implementation of the Turtle Escape Device (TED's) rules, the Florida Department of Natural Resources received a grant from the NMFS to begin a major effort to standardize aerial stranding surveys in cooperation with the established ground network. The primary purpose of this monitoring program is to develop statistically valid numbers of stranded turtles in the NMFS statistical zones 4 and 5 which are located within the subregion. Elsewhere, other targeted zones are part of their program.

The Southwest Florida aerial survey is conducted weekly from Hurricane Pass, Pinellas County, south to Big Marco Pass, Collier County. Should stranded turtles be observed, the appropriate STSSN participant is notified and responds to the stranding.

21
Interactions with Humans

I have been asked frequently by people who have accompanied me on summer night loggerhead turtle patrols or by members of an audience to whom I had just made a presentation: Are loggerhead turtles dangerous, or, do they bite?

Dangerous interactions between loggerhead turtles and humans, although far from common, do occur. I have never personally suffered any traumatic injuries, nor experienced any serious threats from handling the animals over the years, but nonetheless I always insisted on caution whenever coworkers routinely assisted me. Most of my turtle-related injuries have occurred while I was positioning a flipper in which I was about to place a tag. The incredible strength of a turtle will pull the flipper back close to the shell where sharp barnacles can easily cut a hand. Once, one of my young Boy Scout volunteers slipped while turning a loggerhead, and lacerated his chin on barnacles.

Only once have I encountered what can be termed an aggressive female loggerhead turtle. During our attempts to position this particular individual into the apparatus to expedite weighing, she constantly tried to bite. She rather viciously kept reaching out with her mouth agape trying to inflict what I am sure would have been a most severe and painful bite.

There are a few recent newspaper accounts originating in the Florida Keys of scuba divers being approached or even grasped by sexually aroused male loggerhead turtles. This is an extremely rare occurrence, but one that would be overwhelming and frightening. I can imagine how difficult it would be to extricate oneself from the amorous embrace of a three hundred pound loggerhead turtle.

Outside of the subregion, there exists a little understood threatening relationship between marine turtles and man. Outbreaks of turtle poisoning, or Chelonitoxication, infrequently are reported from tropical areas of the earth. It occurs when people consume the flesh of a sea turtle infected by a toxin. The origin of Chelonitoxin is unknown, but it is the opinion of most investigators that the toxin is produced by a poisonous marine algae that is eaten by sea turtles. Symptoms of turtle poisoning are similar to those of ciguatera, a malady that sometimes results after human consumption of some reef fishes. Outbreaks are spotty in distribution. A particular species of sea turtle may be safe in one locality yet, in another adjacent area, eating the flesh of the same form may be deadly. Most recorded cases have occurred in the Indian and Pacific Ocean regions, but the ailment is also known in the Caribbean. Turtles of the genera *Chelonia*, *Eretmochelys*, *Caretta*, and *Dermochelys* have transmitted this alimentary toxicosis. One hundred thirty persons in Quilon, Kerala, India, were affected after eating the meat of hawksbill turtles and eighteen of those victims succumbed to the symptoms. The human fatality rate from Chelonitoxin is approximately 28 percent.

22
Protection of Sea Turtles in Florida

Taking of marine turtles from North American nesting beaches and coastal waters for food or use of the byproducts has been standard practice since the first band of native Americans arrived at the continent's southeastern seashores. Loggerhead and green turtles were hunted on the beaches during the summertime and fished for with special nets or implements during other seasons. When Europeans colonized areas in the region, they established a long-standing traditional use— the harvest of female sea turtles from beaches during their reproductive visits and the collection of their eggs. At first, subsistence take in itself did little damage to the Nation's sea turtle stocks; however, commercialization soon moved into the picture and, because of increased harvesting, some populations were ultimately doomed. Sea turtles, especially green turtles which are savored for their flesh, were being harvested from Florida waters in increasing quantities. It seems, however, that in the past decade the Florida population of nesting green turtles has started to increase. More green turtles are probably the result of long-term protection, active enforcement, and also the change in the cultural composition of the State's human population. Public awareness, through the educational efforts of organizations with goals similar to Caretta Research, Inc., have definitely played an important role in the reversal of the decline of green turtles in Florida.

The Florida Legislature first approved statutory protection for loggerhead and green turtles on "the coasts of the State of Florida" on May 25, 1907. This statute, Chapter 5669—Number 74, protected loggerhead and green turtles "while any such turtle is laying or found out of the waters or upon the beaches of the State of Florida during the months of May, June, July, and August of any year." In 1953, regulations were adopted that afforded protection to marine turtle nests or eggs. By 1963, this protection was extended to other sea turtles "while such turtle is on the beaches of Florida or within one half mile seaward from the beaches" during the summer months.

This closed season concept gave Officers of the then Florida State Board of Conservation, the forerunner of today's Florida Marine Patrol, the enforcement authority to apprehend and arrest violators. Knowledge of the closed season law certainly discouraged many people from illegally taking nesting sea turtles from subregional beaches, although each summer, turtles continued to be taken off beaches by individuals who simply would not accept the protective measures aimed at protecting the dwindling resource. This is analogous to what has gone on with shrimpers and TED's for the past decade. They have fought change, refused to readjust their thinking and technology, and in the meantime the sea turtle resource has been their victim.

I remember when I was a teenager, it was customary for members of many of the older Naples families to go turtle turning in the summer. Most of them seldom harvested more than a loggerhead or two over a summer, and they were convinced that their method of taking turtles from the beach was sound and actually served the conservation of the species. Everyone I knew who harvested nesting loggerheads insisted that he always did it the proper way: waiting until the turtle had deposited her eggs before cutting her throat or striking her in the head with an axe. This practice was based on the assumption that if one ensured the female produced eggs before she was dispatched, she would transfer her loss from the population through the clutch of eggs she had left behind in the sand. I now know how wrong they were!

In 1972, a rather bizarre incident occurred in northern Collier County which eventually led to major improvements in sea turtle protection—not only in Florida, but on a national level too. During a routine night patrol, searching for nesting loggerheads on Vanderbilt Beach, Richard Beatty encountered a group of people standing close to a nesting turtle. He stopped his Honda and was preparing to obtain the animal's measurements and apply a tag when four men from the group approached and informed him that once the turtle had finished nesting, and was swimming away from the beach, they were going to kill it. They based their intent, and later action, on Florida law that existed in 1972. A quote from the last paragraph of Richard Beatty's report of the incident summarizes the outcome: "With the aid of a flashlight they led the turtle to the water, and when the front flippers entered the water someone took a large caliber pistol from the boat and shot her in the head. I watched, helpless to do anything because of a ridiculous legislative response to a minor industry in a small community on Florida's Gulf Coast. Until our leaders restore improved regulations which govern the taking of sea turtles on the West Coast of Florida, nesting sea turtles will remain imperilled, year in and year out, and never safe from the hands of fools."

His comments concerning "legislative response to a minor industry" reflected the confused state of affairs sea turtle conservation legislation was in at the time. The dying turtle fishery in the Cedar Keys area, in a last stand effort, had successfully lobbied the State Legislature to retain provisions that allowed them to continue taking sea turtles. The statute, chapter 370, section 370.12, contained four paragraphs which provided levels of protection, depending on the location where a sea turtle was taken, killed, possessed, or mutilated. For a historical perspective these paragraphs are included below:

(b) 1. It is unlawful for any person to take, kill, possess, mutilate, or in any way destroy any loggerhead, trunkback (sic), leatherback, hawksbill, or ridley, or take or possess any part thereof, while such turtle is on the beaches, sand dunes, or territorial waters of the east coast of Florida from the Georgia line through and including Dade County, during the months of May, June, July, and August.

2. It is unlawful for any person at any time to take, kill, possess, mutilate, or in any way destroy any green turtle, or take or possess any part thereof, while such turtle is on the beaches, sand dunes, or territorial waters of the east coast of Florida from the Georgia line through and including Dade County.

3. It is unlawful for any person to take, kill, possess, mutilate, or in any way destroy any turtle, or take or possess any part thereof, while such turtle is on the beaches or sand dunes of the west coast of Florida between Monroe and Collier Counties and the western boundary of the state, *excepting turtles in the territorial waters thereof having a carapace measurement of more than twenty-six inches.*

4. It is unlawful for any person to take, kill possess, mutilate, or in anyway destroy any green turtle, or take or possess any part thereof, while such turtle has a carapace measurement of not more than forty-one inches unless such person has a certified invoice that said turtle was shipped into Monroe County from a foreign country or outside the territorial waters of the State.

The italics addition to the section of sentence in paragraph 3 is mine. It is this identified exception that provided, at least from the viewpoint of those persons who perpetrated the taking, the loophole by which the turtle was killed and taken in the presence of Richard Beatty.

We submitted this evidence to Federal, State, County and local officials and to every major conservation organization nationwide, urging immediate changes in Florida's sea turtle protection statues, and inclusion of the loggerhead turtle on the Federal Endangered Species List. Earlier, in 1970, the Kemp's ridley, hawksbill, and leatherback turtles had been listed as Endangered by the U. S. Fish and Wildlife Service.

Finally, after considerable encouragement and aggressive pursuit for the listing of the loggerhead turtle, in cooperation with other sea turtle groups and individuals, the loggerhead and green turtles were listed on July 28, 1978. The loggerhead is included as Threatened and the Florida population of the green turtle is listed as Endangered. These categories identify levels of risk for wild flora and fauna. An endangered species is in danger of extinction in all or part of its range, while a threatened species is likely to soon become an endangered species in all or a section of its range.

The Endangered Species Act of 1973, Public Law 93-205; 87 Stat. 884, became effective on December 28, 1973, and has since been reauthorized by Congress. The goals of the Act are aimed at providing a means whereby the ecosystems on which endangered and threatened species are dependent can be conserved, and will be provided a program for the conservation of these species. Further, relative to Endangered Species—it is unlawful for any person(s) subject to the jurisdiction of the United States to import, export, sell or ship in interstate or foreign commerce, harass, harm, or capture, any such species, within the United States and its territorial sea, or on the high seas. For Threatened Species, it is unlawful for any such person(s) to violate any promulgated regulation pertaining to such species. All of these restrictions are applicable to a live or dead specimen, to any of their parts, and to any product made from or including parts of a specimen.

Today, the State of Florida also provides a highly refined and improved piece of legislation which protects marine turtle resources. This revised statute now reads:

MARINE TURTLES PROTECTION ACT

Section 370.12, Subsection 1, Florida Statutes

(1) PROTECTION OF MARINE TURTLES, NESTS AND EGGS: PENALTY.

(a) No person may take, possess, disturb, mutilate, destroy, cause to be destroyed, sell, offer for sale, transfer, molest, or harass any marine turtle nest or eggs at anytime.

(b) No person, firm, or corporation shall take, kill, disturb, mutilate, molest, harass, or destroy any marine turtle, unless by accident in the course of normal fishing activities. Any turtle accidentally caught will be returned alive to the water immediately.

(c) No person, firm, or corporation may possess any marine turtle or parts thereof unless they are in possession of an invoice

> evidencing the fact that said marine turtle or parts thereof has been imported from a foreign country or outside the territorial waters of the state, or are possessed under special permit from the Division of Marine Resources for scientific, educational, or exhibitional purposes.

Violation of any of the above paragraphs is a misdemeanor of the first degree.

Sea turtles cannot be safeguarded by action of lawmakers simply passing these protective regulations. Such legislation is not a cure-all prescription that in any way, shape, or form guarantees survival of these stressed species. Even if coupled with excellent enforcement, the best of legal aid falls short in benefits to their populations. Another critical issue of protection essential to their long-term survival, is public acquisition of their nesting habitat. This means that what little sea turtle nesting beach remains undeveloped, or even beachfront with minimal development in place, must be somehow acquired, and soon. The status of beach purchase should become a much higher priority, internally and at all levels, for each agency involved in land stewardship and resource management. Those staff people who develop land acquisition budgets for such agencies should become better educated, be taught to recognize the significant need, and push for such a program. In turn, Legislators at all levels must become convinced that acquisition of beachfront lands, which can be managed for marine turtle population enhancement and as publicly-owned open space with recreational values, is essential. From the national viewpoint, agencies and Congress must come to realize that there is much at stake and the term wildlife means more than ducks. Land acquisition should not be programmed just for purchase of wetlands.

Along the coast of the subregion, there are approximately eight State Parks which, because of their location, have protected some of the barrier islands from the threat of future development. The beaches which attract loggerhead turtles to their nesting rookery at Cape Sable, in Everglades National Park, are also safe from exploitation. The probable acquisition, by the Federal government of more land in Collier County, north of Everglades National Park, will result in the establishment of the Ten Thousand Islands National Wildlife Refuge. This positive action will help to protect the small outer keys where *Caretta* comes ashore to nest in a primeval environment, not unlike that in the National Park. There is also a move underway, by national conservation groups and sea turtle conservationists, urging the Federal, State, and County governments to begin a bold land acquisition program in Brevard County. If this becomes a reality, the program shall result in a twenty mile section of prime sea turtle beach receiving optimum protection. Further, as a tribute to the late marine turtle scientist, Archie Carr, it has been suggested that the Archie F. Carr National Wildlife Refuge be established once the land acquisition program begins. Federal funding for this important acquisition may be available from specially appropriated monies which originate in the Land and Water Conservation Fund. Congress must act if funds are to be expended from this Fund which was established for purchase of lands for outdoor recreation as well as protection of endangered and threatened species.

Smaller park units, which are county or city owned, dot the barrier island beachfront on the Southwest Florida coast. Unfortunately, most of these are not managed with sea turtles in mind. On Sanibel Island, Lee County owns and operates the Bowman's Beach Regional Park. This wild beachfront is nearly three miles long and is being used by increasing numbers of recreative humans in the daytime; but, at night, parts of it become transformed into one of the better loggerhead turtle nesting beaches in the subregion. Frankly, I am concerned over what might someday happen to parts of this beach. Lee County has a reputation of the left hand not knowing what the right hand is doing and, seemingly, is unable to plan far enough ahead. There has been some recent lip service by Lee County officials to the effect that they will initiate a *Casuarina* removal program on this beach, but from my point of view this plan is not designed to effectively improve nesting habitat for sea turtles. It is my understanding that the County intends to cut down much of the *Casuarina* stand that is invading the beach ridge. This is a futile means of improving nesting habitat, for the stumps with their root systems remain intact to interfere with egg cavity excavation by loggerheads. In my opinion, it would be more cost effective, and more conducive to nesting habitat, to jerk the trees, roots and all, out of the ground with heavy equipment, stockpile and later burn them. Another disadvantage of leaving *Casuarina* remnants in the beach face is that in dynamic beach situations, like Bowman's Beach which changes its profile almost daily, the remains of the trees will soon tumble into the water creating a morass of dead stumps, not the vista of beach which the County is attempting to provide.

A few years ago, when I brought the problem which *Casuarina* poses to their attention, the Sanibel City Council agreed to have the City's Public Works Department remove invading *Casuarina* along the foredune of their one thousand foot long Gulfside City Park. The City, unlike Lee County, has remained very sensitive to the needs of wildlife and their habitats. Over one hundred medium sized Australian pines were *pulled* out, roots and all. Today, there are no *Casuarina* growing in the beach zone of this park which is utilized by loggerheads, and the turtle nesting frequency has increased there.

Back in 1969, Bill Hammond, Director of Environmental Education for the Lee County School System, and I were successful in having the Lee

County Commissioners, in office at the time, agree by resolution to designate most of today's Bowman's Beach as a sea turtle sanctuary. This resolution discourages the use of artificial lights on this beach during the loggerhead turtle nesting and hatching season.

To bring a halt to ongoing degradation of Florida's beaches, the State Legislature promulgated and established the Coastal Construction Control Line in 1971. This legislation prohibits excavation or removal of beach soils, alterations to dune elevations, damage to dunes or the vegetation growing on them, seaward of this Line. This setback line applies only to the State's shoreline that is exposed directly to the Gulf of Mexico or the Atlantic Ocean.

Florida's Coastal Construction Control Line was a quantum leap forward for protection to barrier islands and beaches and the associated benefits to the nesting habitat of sea turtles. No longer are concrete sheet pilings allowed to be inserted in the foredune system, nor are dunes or important plant communities scraped away to level construction sites. Setting precedence in the subregion, the City of Sanibel utilized the setback boundary for beachfront development standards early-on and incorporated the statutory requirements into the language of the City's Comprehensive Land Use Plan adopted in 1976.

One aspect of protection, in the broadest sense, has been captive rearing, or headstarting hatchling sea turtles. Headstarting procedures were first developed and implemented in hopes of enhancing the survival rates of endangered and threatened marine turtle species. The maintenance of hatchling sea turtles in captivity for their first year of life may have eliminated some pelagic predator threats, at least during their very early life history; however, there have been questions raised as to the long-term survival potential and the reproductive future of such animals.

In 1969 and 1970 Caretta Research, Inc. operated a headstart facility on the shores of Tarpon Bay, Sanibel Island. Hundreds of hatchling loggerheads which originated from eggs which had been transplanted into the Lighthouse hatchery compound were reared in holding tanks. The operation was plagued with problems and the program was discontinued in 1971. Vandalism, cold winter water temperatures, and disease were the major obstacles to the success of the operation. Turtles were tagged and released after about a year in captivity. One of these headstarted loggerheads was released in the Gulf at Indian Rocks Beach, in Pinellas County, and was recaptured fourteen months later, apparently in good health, off the coast of Louisiana.

Today, conservationists agree there is no need for such captive-rearing of loggerhead turtles and those that are maintained in captivity are done so for educational, scientific, or rehabilitation purposes. Time and money is better spent protecting nesting adults and their eggs on the beaches. Little loggerheads provided valuable experience and data in the development of techniques which are still used in sea turtle husbandry in some parts of the world. The Florida Department of Natural Resources headstarted green turtles for decades until the program was phased out in 1988. The National Marine Fisheries Service continues to conduct an expensive headstart operation for the Kemp's ridley at their Galveston Laboratory in Texas.

A loggerhead, headstarted at the Kemp's ridley rearing facility, became a world traveler. This loggerhead was released with fifteen hundred yearling ridleys in the Gulf of Mexico fifteen miles off Padre Island, near Corpus Christi, Texas, in June, 1982. In August, 1986, just over four years later, this turtle was captured by a swordfish fisherman on the southeastern coast of Italy, in the Adriatic Sea.

The epitome of headstarting, which, however, is not conducted in the framework of a population enhancing effort, exists on Grand Cayman Island. There the Cayman Turtle Farm raises green turtles to fill demands for green turtle meat and a variety of byproducts. However, the Grand Cayman government-owned firm is not self-sufficient for it continues to import wild-produced eggs from South American beaches. A relatively small number of eggs originate from captive-breeding adults, but in numbers insufficient to meet the trade demand. Presently, their markets for green turtle products are primarily restricted to certain European countries and Japan. The Endangered Species Act precludes importation of even captive-produced sea turtle merchandise into the United States or its territories. Periodically, there are attempts by the Caymanians to have this stipulation of the Act eliminated to permit importation of green turtles produced in the Island's turtle mariculture program, but those in this country responsible for implementation and enforcement of the Act have remained steadfast—so far.

Additional protection of sea turtle populations in Florida gained momentum in the eighties. State statutes and related administrative codes protecting marine turtles have remained effective tools and the Endangered Species Act was reauthorized by Congress. Many of Florida's coastal counties passed or started development of sea turtle protection ordinances. These ordinances are designed, for the most part, to control levels of artificial illumination that reach the beach nesting zone utilized by reproducing sea turtles. These regulations should provide a general reduction in the instances of hatchling disorientation as the little turtles reach the beach surface and call on age-old sea finding cues.

23
Reflections

In retrospect, as I look back over these many years, my work with loggerhead turtles has been exciting and personally fulfilling. Even if I could, I would not make any changes whatsoever to this aspect of my life.

I have witnessed a great blooming of interest in sea turtles and their conservation. When I started out, I worked alone and loggerheads were paid little attention to, at least by people other than those who traditionally harvested the defenseless creatures on the beaches. Law enforcement was lax because most of the Officers of the Florida State Board of Conservation stationed in Southwest Florida, whom I knew personally, had been hired locally and were not professionals. Some of these men sympathized with their neighbors, relatives, and friends who harvested loggerhead turtles. A few of these early law enforcement officers even went so far as to overlook the taking of one loggerhead turtle per season, by a family. After all, many of those early officers had been raised to turn turtles as part of their heritage.

Today, hundreds of people have been mustered to help sea turtles in the Southeastern United States. Their conservation has become a popular cause in this country, and officers of the modern Florida Marine Patrol are highly trained, dedicated, and professional men and women. This is good, for these endangered and threatened marine animals will continue to need all the help they can get. However, the nonenforcement aspects of sea turtle conservation must never be allowed to reach a plateau where it becomes just the "in" thing to do. People who become involved in the program must make a commitment, remain steadfast, and forge ahead. I have known many individuals who joined the ranks with enthusiasm, but when the work became too routine for them or somehow conflicted with other facets of their personal lives, or when the night-flying biting insects became unbearable, for man or turtle, they defected from the cause. I have been fortunate enough to have been surrounded by a core of very dear friends who have shared in my sea turtle efforts and commitment. When things were not going exactly right, or at times when I approached exhaustion, they revitalized my enthusiasm and kept me on course. They made personal sacrifices too, but like me they persevered and continued to do good deeds for loggerhead turtles. Like old soldiers, we real turtle people never die, we just fade away. As I begin to enter the fade away phase of my life, I do so knowing that what I started so many years ago shall continue and grow, nurtured by people who care.

In my early years in this field, conditions were simply deplorable for loggerheads and there was no one else out there to come to their rescue. It wasn't until the late fifties that scientists were just beginning to take a real interest in the marine turtle group. Even then, no one comprehended the big picture in terms of decline of certain nesting populations, or what the not too distant future had in store for sea turtles—except perhaps Archie Carr.

My attitude, back so many years ago, toward what was happening was based on personal observations which to me, even as a teenager, were obvious threats to the population, and I must say circumstances were very depressing. My perspective at the time was best expressed in the lines of some verse I wrote as a young man and I will share these and leave you with them.

The Loggerhead's Legacy

Imagine a starry night in the month of June,
with an island beach outlined by full moon.
This is the time loggerhead turtles creep
on land to nest and return to the deep.

A head breaks surf and looks towards shore,
as her ancestors did ages before.
The strange force which compels her so
is her legacy, but she does not know.

She leaves the surging endless tide,
eyes scanning the brilliant beach so wide,
studded with seashells and pockets of foam,
the first time she has left her safe sea home.

Though dangers are many, instinct is strong,
no fear daunts her efforts, however long.
She seeks the safest burial site,
to deposit her eggs in secret at night.

Slowly she struggles up the sand slope,
bearing within her the continuing hope
of a new generation, like her, who shall be
innocent and unwarned when they leave the sea.

Foot by foot she clambers the beach
to grass-covered dunes beyond the tide's reach.
Her body weighs heavy, she stops to lie,
To rest a moment and give a great sigh.

This annual struggle beneath the summer moon
is an eternal battle with elements, man and 'coon.
She is watched by eyes gleaming with hate
as human and animal plan her sad fate.

Careful, unknowing, she tests the sand,
on this her first journey to the land
since she herself escaped the shell,
left behind the beach, in the sea to dwell.

Each foot probes for a nesting place
as she seeks to guarantee her race
a continuing place in Nature's great scheme,
a cycle of reality and not a lost dream.

Slowly, precisely she lifts the sand
with flippers sensitive as a human hand.
Scoop after scoop the cavity grows wide,
and deepens, her precious eggs to hide.

From sparkling eyes tears start to flow,
but not from emotional reasons we know.
Although salt and sand settle in each eye,
bodily functions cause her to cry.

With chin pressed into the sandy earth
she begins the labor of giving birth.
Into the cavity the groups of eggs drop,
by twos and fours, at two hundred they stop.

Her mission completed, hind feet work to cover
nest and eggs wild thieves wait to discover.
With wing-like forelegs she scatters sand neatly,
to camouflage her nest site completely.

Weary from effort she heads for the surf,
leaving obvious tracks in the sandy turf.
A beam of light flashes on the sand,
flippers touch water, but she is still on land.

Eager hands grasp at her shell.
Hands of fools, contrived in Hell.
Human beings with no shame
who think their actions only a game.

Unmercifully, she is thrown upon her back,
beside her sea home, above her track.
On a bare steel blade the moonlight flashes,
from its thrust her life's blood dashes.

From this creature of God, ever so humble
comes a last sigh and a pitiful rumble,
as she and her life are cast aside,
no more to play in, or race against the tide.

The lament of the ancient loggerhead,
another poacher, another sea turtle dead.
Man has influenced those who live in the sea,
and left the loggerhead this tragic legacy.

Appendix

Allied Species in the Eastern Gulf of Mexico

The loggerhead turtle occupies the waters of the subregion in the company of four related species. Some of these forms occur regularly along the coast of Southwest Florida and others are rarely observed. As the text and format of this book evolved, it became increasingly apparent to me that I should include some commentary on the loggerhead's kin. My records and data sheets are voluminous, representing statistics on thousands of sea turtles, including a number of individuals of the four allied species. I have added this appendix to discuss each of the relatives of the loggerhead turtle which also inhabit the eastern Gulf of Mexico.

I have utilized a field guide format, with emphasis on the respective species' description and general biology. Following each of these technical segments I include a remarks section in which is discussed interesting aspects of their life history and documentation of their presence in the subregion, based on stranding records or other evidence.

Chelonia mydas

CHELONIA MYDAS

Scientific name: *Chelonia mydas* (Linnaeus, 1758)

Glossarial of above: (chelonia), tortoise; (mydas), wet

Vernacular name(s), (U.S.): Green turtle

Identification: The carapace has four pairs of costals and the carapacial scutes do not overlap. Identifying head scalation characters consist of a single pair of prefrontals. Normally, there is one claw on each foreflipper. Straightline carapace length to 60 inches (153 cm). Color is variable, light grayish green to olive black with radiated or spotted flecks sometimes present over the carapace. The plastron is a light amber yellow in adults.

World-wide distribution: *Chelonia mydas* is a cosmopolitan species which historically congregates in large colonies on mainland beaches and sometimes on oceanic islands, i.e., Ascension Island. Generally, their nesting distribution is situated between 30 degrees south latitude and 30 degrees north latitude.

Gulf of Mexico distribution: The green turtle, juveniles and adults, occurs along the shoreline of the eastern Gulf.

Eastern Gulf of Mexico population levels: Instances of incidental catch by inland waters commercial fishermen have decreased over the past decade. No accurate population estimate is available. Elsewhere in Florida, the nesting population appears to be increasing. An estimated 500 nests were deposited by *Chelonia* along the Florida Atlantic Coast during 1989.

Habitat: Shoal marine waters having well vegetated bottoms.

Diet: Adult green turtles are herbivorous, feeding on such marine grasses as eelgrass *(Zostera)*, manatee grass *(Syringodium)*, turtle grass *(Thallasia)*, and seagrass *(Halophila)*.

Terrestrial gait: While on the land for nesting, the green turtle literally pulls itself around. Simultaneous outreaching of the front flippers, resulting in a lunging motion, moves the animal forward. This technique of locomotion creates a nesting crawl configuration of opposite fore-flipper imprints on the beach surface.

Nesting period: In Florida, nesting occurs during the summer months and appears to peak in July.

Nesting periodicity: Frequency of reproductive seasons is variable for *Chelonia mydas*. Various populations around the world contain individuals that nest at two, three, and four year intervals. In general terms, green turtle nesting periodicity probably averages three years; however, on the east coast of Florida a two year cycle is most frequent.

Nest site preparation: Once site selection is made, the green turtle excavates a body pit that, in effect, lowers the animal below the relative adjacent beach grade. Formation of this depression permits removal of several inches of dry topsoil and gains additional depth for the egg cavity.

Nest construction: Alternate probing and curling action with the distal edge of the rear flippers removes soil at a steady pace from the egg chamber. Dimensions of the pear or flask-shaped cavity are variable and dependent on the size of the turtle.

Oviposition: Once nest construction is finalized, a nesting green turtle will relax and position the rear flippers in such a fashion as to cover the egg chamber opening. As described earlier for *Caretta*, other members of the Cheloniidae place their rear flippers more offset, or on either side of the nest hole. During actual oviposition there is little noticeable movement of the adult.

Egg size: The average egg size for *Chelonia mydas* in Florida is 46 mm.

Clutch size: The number of eggs produced per clutch is highly variable. The range for green turtle egg complements is 80 to 226, with an average for Florida females of 145.

Internesting interval and multiple nesting: Eleven egg clutches in one nesting season is the record for this species, and were produced by an individual *Chelonia mydas* in Sarawak, Malaysia. The great majority of green turtles do not even approach this many nests in one season. Intervals between nesting emergences range from 11 to 15 nights. In 1985, sea turtle biologists documented five nestings for a North Carolina green turtle. This specimen deposited eggs on the average of every 11 nights. Incidentally, it was reported that this same individual deposited a total of 893 eggs in the five nestings. Nesting by *Chelonia* so far north in the United States was a very rare occurrence.

Incubation period: Several factors regulate the time involved in the development of marine turtle eggs. The incubation time frame, from deposition to hatchling emergence, ranges from 45 to 75 days.

Hatchling size and identification: Typical green turtle hatchlings from Florida are about 50 mm in straightline carapace length, 40 mm in width, and weigh about 25 grams. Hatchling green turtles are quite easy to identify since they are the only young sea turtles in the Atlantic Ocean system that have a snow white plastron. Other identifying characters include the scalation configuration of the adult.

Remarks: Small green turtles, in the ten to twenty pound range, are occasionally netted in the inland marine system along the Florida Gulf Coast by bait shrimpers and commercial net fishermen. Occasionally, cold-stunned subadults are discovered in inland waters during periods of low water temperatures in the winter.

Sexually mature green turtles are rarely observed in the littoral waters of the eastern Gulf of Mexico. A review of over three decades of dead sea turtle stranding data for Sanibel Island documents only two records of adult green turtles. Although infrequent, strandings of subadults, are much more common.

In October, 1959, a live, female green turtle, which appeared to be healthy, was discovered early one morning high on the beach immediately to the west of the Sanibel Lighthouse. I observed and measured the individual after its discovery by Ken Havourd, a longtime Sanibel resident. The turtle was 39 inches long over the curve of the carapace. Because of State regulations existing at the time, the turtle was legally removed by Havourd despite my request that it not be taken.

I never determined just why this animal was on the beach. It had moved a considerable distance from the water, well over fifty feet, and had entered the dense sea oat fringe. It is very possible that this green turtle had moved ashore to nest and was discovered and taken before it had the opportunity.

Years later, in June 1976, a decapitated but otherwise intact adult female measuring 37 inches, curved carapace length, was found stranded on Sanibel's Gulf beach ten miles west of the Lighthouse.

Older residents living on the barrier islands of Collier, Lee and Charlotte Counties informed me many years ago that green turtles were occasionally taken by them, for food, from the Gulf beaches in the early part of this century. As late as 1967 an adult nesting green turtle is purported to have been killed on the Collier County section of Bonita Beach by poachers.

Historically, green turtle nesting on the coast of Southwest Florida probably has been very rare. The species prefers high energy beaches which have a higher profile than is typically available on the lower Florida Gulf coast. Because of the standard deep body pit, nest excavation depths require higher dune formations. An elevated beach provides drainage of rain or tidal water from the nest, and this is of paramount importance for successful incubation. In 1986, 1987, and again in 1988, green turtles nested in Monroe County on lands of the Key West National Wildlife Refuge according to Deborah Holle, Manager of the Refuge. In 1986, an unsuccessful egg clutch was deposited on an unnamed island in the Marqueses Keys, about twenty miles west of Key West. In 1987, a green turtle nest was located on Boca Grande Key, some twenty miles southwest of Key West and was partially successful, producing an unknown number of hatchlings. In 1988, another green turtle visited and nested on an island of the Marqueses Keys. This nest contained 127 eggs, of which 114 hatched. The islands of the Dry Tortugas group, once a green turtle nesting area, have been the location of some green turtle nesting in the mid-1980's according to biologists at Everglades National Park.

In the mid-1970's, Caretta Research, Inc., in cooperation with the Florida Department of Natural Resources, headstarted scores of hatchling green turtles on Sanibel Island. After eight to ten months in captivity, the turtles were tagged and released into the Gulf of Mexico. To my knowledge none of the tagged green turtles which were raised and released from Sanibel were ever recovered.

A strange skin malady which occurs in *Chelonia mydas,* was first reported nearly fifty years ago. The affliction has increased in frequency in recent years, and has become a serious threat to the future of this species. Tumor-like growths, known as green turtle fibropapillomas, develop on the skin and near body openings such as the eyes, nostrils and mouth. The tumors may grow to weigh pounds and impede the animal's movements and ability to feed. At least two of the stranded juvenile green turtle carcasses which I have examined over the years on Sanibel Island had been the direct result of starvation caused by the presence of near fist-sized growths which completely covered each eye, making it impossible for the turtles to find food.

The tumors are believed to be viral in origin, but veterinarians have been unsuccessful so far in isolating the responsible virus. Even if a cause is determined soon, it will be impossible to develop and provide treatment to wild green turtle populations. These tumors seriously are impacting green turtles in Florida and the Hawaiian Islands; however, the infection is not restricted to these regions and has been recorded in many other parts of the world. It has been reported that perhaps half of the green turtles which dwell in the Indian River lagoon system of Florida's Atlantic coast are carrying these grotesque tumors.

The outlook for halting this life-threatening ailment is remote indeed—as if this endangered species doesn't already have enough trouble surviving wherever it occurs!

LEPIDOCHELYS KEMPI

Scientific name: *Lepidochelys kempi* (Garman, 1880)

Glossarial of above: (lepidochelys), scaly turtle; (kempi), proper name of R.M. Kemp.

Vernacular name(s), (U.S.): Kemp's ridley, Atlantic ridley, Gulf ridley, Bastard turtle

Identification: The carapace has five pairs of costals and the scutes do not overlap. Two pairs of prefrontals are present and the plastron is unique in that a pore is located on the rear of each of the four (sometimes three) inframarginals. The carapace is gray in juveniles and grayish olive-green in adults. Straightline carapace length to 27 inches (70 cm). The Kemp's ridley is the smallest of the marine turtles. This is the only sea turtle in the Gulf of Mexico where, on casual examination from above, the shell appears to be as wide as it is long.

World-wide distribution: Restricted for the most part to the Western North Atlantic Ocean. The population center is the northern Gulf of Mexico, rarely to the upper Caribbean, and frequently along the U.S. Atlantic coast to Nova Scotia. There is at least one recent non-nesting capture record of the Kemp's ridley from the Mediterranean Sea.

Gulf of Mexico distribution: The Kemp's ridley is distributed along the entire coastline of the Gulf of Mexico but concentrates for nesting on Mexican beaches in Tamaulipas State, north of Vera Cruz. Occasionally, lone females have nested on Padre Island, Texas.

Eastern Gulf of Mexico population levels: A depleted population exists in the subregion. Until the early 1970's, many small ridleys were netted by commercial mullet fishermen, but today this appears to be a rarity. I personally have not heard of a ridley being caught by gillnetters around the Sanibel-Captiva area since 1976, but they are occasionally caught by shrimp trawlers in the littoral waters of Southwest Florida.

Habitat: Preferred habitat in the subregion appears to be waters of the open Gulf of Mexico, coastal sounds and bays.

Diet: The Kemp's ridley is fundamentally a marine carnivore. An extremely limited number of dead ridleys have been available to me for necropsy and examination of their gastrointestinal contents. Remains of the following crustaceans have been observed: blue crabs *(Callinectes sapidus)*, spider crabs *(Libinia dubia)*, decorator crabs *(Stenocionops furcata)*, and stone crabs *(Menippe mercenaria)*.

Terrestrial gait: The flippers are used alternately as the ridley moves across land. Because of the animal's relatively small size and light weight,

Lepidochelys kempi

the flipper imprints are shallow and the footfalls result in an alternating foreflipper pattern.

Nesting period: Nesting occurs from April through July, it is sporadic and its timing may be weather-related. Female Kemp's ridleys usually arrive on the nesting beach during periods when onshore winds are strong and the surf is rolling well up the beach.

Nesting periodicity: According to Mexican ridley researchers, 58 percent of Kemp's ridleys nest in successive years, 29 percent every two years, and 13 percent every three years.

Nest site preparation: During beach ascent, this species of sea turtle frequently moves forward with periodic pauses and what appears to be insertion of the lower jaw into the sand. As discussed earlier, there is some question as to the reason, or role, of this behavior. No body pit is prepared; the only action prior to excavation of an egg chamber may be foreflipper sweeps to remove some dry sand on the beach surface. Thereafter, the front flippers are pressed firmly into the sand, apparently as an anchoring technique.

Nest construction: The posterior flippers are utilized alternately, their outer edges curling slightly to cup sand for withdrawal.

Oviposition: The egg laying procedure is typical sea turtle behavior, but unlike *Chelonia*, the nest opening is not concealed from above by the rear flippers. Eggs drop into the ground in variable quantities, one to four together, until the entire complement is in place. The postnesting and final covering behavior is quite unique. Following oviposition and back-filling the egg cavity, a female ridley will violently sway her body from side to side and alternately thump the borders of each side of her shell against the beach surface.

Egg size: The average egg diameter for *Lepidochelys kempi* is 38.9 mm.

Clutch size: The range of total clutch size is rather broad for such a small turtle, 54 to 185, while the average clutch size for the species is 105 eggs.

Internesting interval and multiple nesting: Data indicate that the Kemp's ridley may nest a maximum of four times per season. There may not be a consistent frequency of multiple nesting emergences, or a regulated internesting interval, in this species. Multiple nestings are known to occur over a range of 10 and 28 days.

Incubation period: The incubation period, the time lapse between egg deposition and hatchling emergence at the beach surface, ranges from 50 to 70 days.

Hatchling size and identification: Straightline carapace length, 38-46 mm; straightline width, 30-40 mm. Generally, the hatchlings are very dark gray to black in color above and below, and the carapace and flippers are edged with white.

Remarks: For many years, sea turtle biologist Archie Carr searched for the nesting beaches of this sea turtle. Florida Gulf coast fishermen commonly called the Kemp's ridley the bastard turtle for they believed that this species was a hybrid between the green and loggerhead turtles. Carr's persistent investigation into the basic life history of the Kemp's ridley finally disproved this belief, but I still hear the term used by some of the older fishermen. Although rare, hybridization among marine turtles of the Cheloniidae does occur.

Studies by Carr and his students revealed that subadult ridleys, and an occasional eggbearing adult female were taken by turtlers in the Crystal River—Cedar Keys area of Florida. Despite the relative abundance of the species in those days, the nesting beach and general reproductive behavior of the Kemp's ridley remained a puzzle. The scientific community was astonished, when in 1963, Henry Hildebrand presented a film during a professional meeting of the American Society of Ichthyologists and Herpetologists in Austin, Texas. The movie, filmed by Andres Herrera in 1947, revealed about forty thousand Kemp's ridleys ashore and nesting during daylight on a remote, little-visited, section of the Mexican Gulf coast in Tamaulipas, south of Brownsville, Texas. This unusual daytime nesting assemblage, known as an arribada, has since become recognized as typical reproductive behavior for this genus. Only *Lepidochelys* regularly leaves the water for nesting during daylight hours, and usually in sizeable groups.

Today, nesting assemblages of this ridley still occur, but with significantly fewer numbers of turtles. A maximum of 500 female ridleys continue to use their ancestral beach each year. Harvest of eggs and adults for human consumption, the leather trade, predators, hurricanes, pollution, and incidental catch by the seafood industry have taken their sad toll.

Headstart projects, the most recent of which is operated by the U.S. National Marine Fisheries Service, are probably assisting the stock in terms of increasing short-term survival. Headstarted turtles may better cope with the dangers of their environment, but the technique will be an uncertain management tool until recognizable female ridleys begin to show up someday on their ancestral nesting beach, or an adopted beach. Unless fishing technology improves, strict enforcement and compliance of protective regulations are adhered to, and egg loss because of predation is mitigated, in both the United States and Mexico the overall survival outlook for the Kemp's ridley is gloomy.

A conservation technique that has been implemented, on a smaller scale, is the translocation of ridley eggs from Mexico to the beaches of the United States on Padre Island, Texas. In recent years, a handful of ridleys have nested on the "adopted" beaches. These may have been produced on the Texas beach by turtles which hatched from eggs that had been relocated years before. Should a nesting colony be established from this effort, it certainly will be advantageous to the long-term survival of this seriously jeopardized sea turtle.

This, the rarest and most endangered of marine turtles, has declined dramatically. Until the middle 1970's, small ridleys frequently were caught by commercial fishermen in most of the sounds and bays along the Florida Gulf Coast. At one time, a commercial ridley fishery was centered near Crystal River, Florida.

In 1988, captive ridleys held at the Clearwater Marine Science Center mated, and a female later moved up an artificial beach and deposited a clutch of eggs. Many of these proved to be infertile, or otherwise unsuccessful, but one hatchling which was conceived and hatched in captivity survived.

On the morning of May 30, 1989, the next to the impossible happened! A gravid Kemp's ridley ventured ashore, between 900 and 1000 hours, on Madeira Beach, one mile north of John's Pass, Pinellas County, and deposited a group of eggs. The turtle was observed nesting by Tony Lozon, a sea turtle permittee of the Florida Department of Natural Resources, and videotaped by a tourist. According to Colleen Coogan, a sea turtle biologist assigned to the Florida Marine Research Institute, a copy of the tape was given to her laboratory and it verifies the identity of the animal.

In early June, storm tides inundated the nest site and the eggs had to be relocated. Pat Castaneda, also with the state laboratory, helped to move the clutch of 106 eggs. She estimated that up to 50 percent of the eggs had been destroyed because of flooding. On July 27, eighteen Kemp's ridley hatchlings emerged from the nest which was surrounded by about a hundred people on hand for the event. Total production for this nest was twenty-four neonates, which emerged over a two day interval.

Stranding records for Sanibel Island include data on a few dead Kemp's ridleys. The first of these to be documented, a female, stranded on the Gulf beach, approximately one-half mile west of the Sanibel Lighthouse on December 21, 1971. It was a sizeable individual with a straightline carapace length of 27.5 inches (70cm) and straightline carapace width of 27.125 inches (68.5 cm) respectfully. Its skull dimensions, with the tomium removed, was 20.5 cm in length (premaxillary to the supraoccipital), and 14.6 cm in width (outer extremities of the quadratojugal). The carapacial shell of this specimen is in the possession of George Weymouth and the skull is in the author's personal collection. The referenced skull bones are identified in Figures 18 and 19 on page 26.

The next specimen known to strand, a subadult, was collected from the Sanibel Gulf beach on October 24, 1980. More recently, in the spring of 1987, four subadult ridley carcasses came to rest on the Gulf beach of Sanibel. The extended lapse of time between strandings events for this species on Sanibel Island is interesting. There were no stranded ridleys observed here in 1988; however, in 1989 ten ridleys, mostly subadult, washed ashore on Sanibel from January through April.

Eretmochelys imbricata

ERETMOCHELYS IMBRICATA

Scientific name: *Eretmochelys imbricata* (Linnaeus, 1766)

Glossarial of above: (eretmochelys), oar turtle; (imbricata), overlapping

Vernacular name(s), (U.S.): Hawksbill turtle

Identification: The rough outer edges of the carapacial scutes typically overlap except in very young or old specimens. Four costal scutes are present on each side of the carapace and the head scalation includes two pair of prefrontals. There are two claws on each foreflipper. The jaws of this species are narrowed, compared to the other sea turtle species. This characteristic is responsible for the hawksbill's common name. In adults the carapace is dark green-brown, while small to medium sized animals have the characteristic, and famous, tortoiseshell coloration. The maximum recorded straightline carapace length is 38 inches.

World-wide distribution: The hawksbill occurs in the major oceans, occupying tropical and subtropical waters. The species nests on a variety of mainland and island beaches, but unlike other sea turtles, *Eretmochelys* does not form large nesting concentrations and nesting is well dispersed throughout the animal's range.

Gulf of Mexico distribution: The hawksbill occurs in limited areas of the Gulf of Mexico. It is known to inhabit the Florida Keys and there are a number of reliable stranding records from the northern Gulf. The species has also been recorded from the waters of Pinellas County.

Eastern Gulf of Mexico population levels: Because of the rarity of the hawksbill's occurrence in the waters of the subregion, estimating a population level is difficult.

Habitat: *Eretmochelys* typically frequents shallow water areas adjacent to coral formations or rocky bottoms; however, the species is not restricted to such habitat and has been reported from mangrove systems.

Diet: Encrusting invertebrates which colonize shoal reef waters apparently are preferred food items. Sponges, tunicates, coelenterates, sea urchins, and mollusks are consumed. There are a few reports of the hawksbill eating mangrove parts, such as propagules and leaves.

Terrestrial gait: Locomotion across the beach during nesting landings is accomplished by alternate utilization of the flippers—much in the same way as *Lepidochelys.*

Nesting period: Nesting occurs between April and November, with a peak period in June and July.

Nesting periodicity: Renesting intervals for *Eretmochelys* appear to occur every two or three years. Since nesting emergences by this species in Florida are so rare, no renesting periodicity data has been established in the subregion.

Nest site preparation: Little development of a nest site is accomplished by the hawksbill. Surface sand is removed by random broadcast action of the front and rear flippers prior to start of the egg chamber excavation.

Nest construction: The rear flippers alternately excavate the egg cavity in typical sea turtle fashion.

Oviposition: The front flippers are anchored in the sand and held close to the carapace, and the rear flippers flank the egg cavity.

Egg size: The average egg diameter, obtained from populations on Caribbean beaches where this species nests, is 38 mm.

Clutch size: As with other sea turtles, there is a broad variability in the number of eggs contained in a clutch. Clutch sizes have been reported to range from 53 to 250 eggs. An average hawksbill egg complement, from the Caribbean populations, numbers 161 eggs. Both the maximum clutch size (250) and the mean (161) seem to be very high for a turtle of such moderate adult size.

Internesting interval and multiple nesting: The frequency of within season internesting land visits by *Eretmochelys* are not fully understood and apparently fall within the 12 to 15 day interval which is normal for the majority of sea turtles. A maximum of five nests per season have been reported for the hawksbill.

Incubation period: The average incubation period is 59 days but, like the eggs of its relatives, the combined time for full term development and emergence at the beach surface is quite variable (50 to 70 days).

Hatchling size and identification: Straightline carapace length at hatching ranges between 38 and 55 mm. Hatchling hawksbills are dark brown and have a vertebral keel on the carapace and double plastron ridges. These features, coupled with the presence of the aforementioned four costal scutes, identify hatchling hawksbills.

Remarks: In 1980, there was a report of a hawksbill nesting, resulting in a successful hatch on Longboat Key, Sarasota County. Information of this unusual event was provided to Ross Witham, a sea turtle biologist then attached to Florida's Bureau of Marine Resources, by a citizens' group known as the Longboat Key Turtle Watch. I contacted persons associated with this organization in hopes that I be given the opportunity to inspect the specimen(s) upon which the identification was made. I was informed that verification had been based on a hatchling group that had been released shortly after hatching. Unfortunately, none had been placed in a scientific collection as voucher specimens and apparently none were photographed. Initially, I was extremely skeptical of the identification and thought that perhaps an aberrantly scaled *Caretta* had been misidentified by the Longboat Key group. I remain puzzled as to why there is no scientific documentation, or evidence of such a very rare event. Frankly, I will continue to doubt that a female hawksbill nested on Longboat Key in 1980. In comparison, the Kemp's ridley, which visited and nested on Madeira Beach in 1989, is well documented with videotape and photographs.

There are no stranding records of hawksbills on the beaches of Sanibel-Captiva, but there is a lone stranding record of an *Eretmochelys* carcass reaching a nearby subregional barrier island. According to Barbara Schroeder, Sea Turtle Coordinator for the Florida Department of Natural Resources, a small hawksbill turtle (53 cm, curved carapace length) stranded three quarters of a mile south of Clam Pass, Collier County, on May 23, 1988. The animal was examined and reported to Barbara by Ron Mezich of The Conservancy in Naples. On December 24, 1989, a small hawksbill that was just under a foot long, was recorded by Keith Sinclair stranded near Fort Desoto Park, in Pinellas County. Since State-wide record keeping was initiated on sea turtle strandings in 1980, there have been several strandings of hawksbills in Monroe County which is situated on the southern border of the subregion.

In the late 1950's and early 1960's, hawksbill turtles were reported to be regular residents of a coral formation a few hundred yards off Vanderbilt Beach in northern Collier County—very close to Clam Pass. This information came to me from scuba divers who were familiar with the species in the Florida Keys. Unfortunately, I was never able to substantiate these reports.

Hawksbill turtles do, very rarely, nest on Florida shores on the East coast. The first documented Florida nest was deposited on Juno Beach, Palm Beach County, in August 1959. There have been at least three additional reliable nesting records for the species in Southeastern Florida since then.

Dermochelys coriacea

DERMOCHELYS CORIACEA

Scientific name: *Dermochelys coriacea* (Linnaeus, 1766)

Glossarial of above: (dermochelys), leather tortoise; (coriacea), leathery

Vernacular name(s), (U.S.): Leatherback turtle; Trunkback turtle

Identification: The largest of marine turtles, the leatherback is unique among modern shelled reptiles. The shell alone is the chief factor of identification. This species lacks a bony shell, having in its place a smooth, leather-like, outer skin in which is imbedded a mosaic of small bones. The carapace is ridged with seven elevated keels and the plastron has a series of five similar ridges. The limbs lack claws. The leatherback is basically dark brown-black with randomly located obscure patterns of white or yellow on the head, neck, flippers and carapace. The plastron is predominately white. By weight the leatherback turtle is the largest living reptile with a curved carapace length of up to 101 inches (256.5 cm) and a maximum recorded body weight of 2,016 pounds.

World-wide distribution: The leatherback is a circumglobal species and occupies tropical as well as temperate waters. This turtle regularly ranges farther north and south than the other marine turtles and has been reported from coastal waters of Newfoundland and Argentina.

Gulf of Mexico distribution: *Dermochelys* is a pelagic species and is known to occur in the Gulf. Large concentrations of leatherbacks have been reported occasionally off of the Texas coast, but no such aggregations are known for the subregion.

Eastern Gulf of Mexico population levels: The leatherback rarely is observed in the waters of the eastern Gulf of Mexico and seldom nests on the barrier islands. There is no available estimate of the turtle's population in the subregion.

Habitat: Although considered to be a pelagic species, it has been known to enter the more shallow waters of sounds and large bays.

Diet: The dietary preferences of wild leatherback turtles are unclear. Jellyfish seem to be the chief food item; however, stomachs of stranded specimens also have contained other animals including squid, tunicates, and marine plants.

Terrestrial gait: Like the green turtle, the leatherback moves about while upon land by simultaneous, lunging movement of the front flippers.

Nesting period: In Florida, the leatherbacks which nest along the Atlantic seaboard are known to come ashore in March through August.

Nesting periodicity: Tagging of *Dermochelys* on beaches where the turtles nest in sizeable groups (Surinam and Malaysia) has indicated that renesting intervals are variable and may occur on two and three year cycles.

Nest site preparation: Development of a body pit appears to be a common trait for nesting leatherbacks; however, some populations of the species, inexplicably, do not perform this behavioral act prior to nest excavation.

Nest construction: Excavation of the egg chamber is similar to the sea turtles of the Cheloniidae.

Oviposition: Like the green turtle, the female leatherback covers the opening of the egg cavity with her rear flippers during oviposition.

Egg size: The average size of normal *Dermochelys* eggs is 55 mm in diameter. Leatherback turtles almost always deposit a number of small yolkless eggs along with the normal size eggs. Often, 50 percent of a leatherback clutch may consist of small atypical eggs which can also be present in a variety of shapes; i.e., elongated, dumbbelled, etc.

Clutch size: Clutches of eggs of this species contain between 50 and 170 eggs, with an average clutch of 85, excluding the undersized or otherwise aberrant eggs discussed above.

Internesting interval and multiple nesting: The interval between nesting landings by the leatherback is relatively short. Multiple clutches are usually deposited about ten nights apart and although rare, as many as nine egg complements can be produced in a nesting season.

Incubation period: The time required for embryonic development and hatchling emergence at the surface of the beach can vary between 53 and 74 days.

Hatchling size and identification: Neonate leatherback turtles cannot be confused with hatchlings of other marine turtles, or for that matter, with any other species of turtle. The hatchlings average 62 mm in straightline carapace length and the shell and skin are covered with tiny pliable scales. As the young turtles grow, these scales are lost and replaced with the leather-like skin and shell covering discussed earlier. They are dark, usually black, with white carapacial keels and flipper edges, and the front flippers are extremely long.

Remarks: Specimens of this large marine turtle occasionally are observed offshore in the subregional waters; however, there are no known records of this species nesting on the lower Gulf coast of Florida. The closest that one of these giant turtles has nested to the subregion occurred in 1974, when a leatherback nested on St. Vincent National Wildlife Refuge, in Franklin County, Florida.

One stranding record exists for Sanibel Island, although the animal was collected years before formal record keeping of sea turtle strandings. On August 5, 1943 James Williams of Sanibel, removed the carapace of a leatherback he had found on the beach near the Sanibel Lighthouse. He stored the oily shell near his beach cottage adjacent to the Lighthouse reservation and in 1959, after learning of my interest in sea turtles, gave the shell to me. The carapace measured 54 inches (137.2 cm), curvature measurement and, because of its oily and very odoriferous nature, I too stored it outside in the brush near my residence at the Lighthouse. Unfortunately, Hurricane Donna in 1960, produced a nine foot tidal surge that swept across Sanibel's Point Ybel and carried my leatherback turtle carapace along with it.

In late July, 1988, while on routine turtle patrol, Eve Haverfield and I discovered a most unusual turtle crawl on Bowman's Beach, Sanibel Island. The track was not over two nights old since that was the interval between our beach inspections. The turtle had travelled straight out of the water for about seventy-five feet when its forward progress was halted by a steep erosion escarpment. The animal had then disturbed some surface sand and false crawled, returning to the water in an angled meandering direction, not perpendicular to the beach. This descent track had two concentric 360 degree turns, about twenty feet apart, on the dry beach. The distinct flipper imprints were opposite each other, not alternate as in *Caretta,* and the central area of the track had an unusual ridged appearance. The outer extremities of this crawl averaged 120 cm apart. The widest loggerhead crawl we have measured was just over 113 cm across. The crawl configuration and the associated signs convince me that a leatherback turtle had visited Sanibel; however, the turtle did not revisit either Sanibel or Captiva that summer.

A few nights later, Bob Averill, a Sanibel businessman who had worked on the Earthwatch leatherback project in the U.S. Virgin Islands just a few months before, examined the track. Bob agreed that our unusual crawl had been made by a leatherback turtle.

Glossary

Some of the terms used in the text may be unfamiliar to the reader; therefore, I have included this glossary of conversions and definitions.

CONVERSIONS

TO CONVERT	*INTO*	*MULTIPLY BY*
Inches	Centimeters	2.540
Centimeters	Inches	0.3937
Inches	Millimeters	25.40
Millimeters	Inches	0.03937
Miles	Kilometers	1.609
Pounds	Kilograms	0.4536
Celsius degrees (C)	Fahrenheit degrees (F)	1.8 & add 32
Fahrenheit degrees (F)	Celsius degrees (C)	.5556 & subtract 32

DEFINITIONS

Addled. Rotten.

Albumen. The white of an egg.

Allantois. A thin membrane that develops in reptiles, birds, and some mammals, from the ventral wall of the hindgut.

Ambient. Surrounding

Anaerobic. Living without atmospheric air or free oxygen.

Asphyxiation. To suffocate.

Basking. To warm oneself in the sunlight.

Benthic. Organisms dwelling on the bottom of a water body.

Berm. A narrow flat beach area, at the top or bottom of a slope.

Biogenic. Originating from living organisms.

Bivariate. A term used in statistics, pertaining to two variables.

Blastula. The early developmental stage which follows the morula stage.

Botany Bay Oak. The former trade name of *Casuarina equisetifolia* when this Australian tree was used as a cabinet wood.

Chelonian. Pertaining to members of the order Chelonia—turtles.

Ciliary. Having short, hair-like cellular outgrowths, which move rhythmically and transport fluids; i.e., eggs through the oviducts in turtles.

Cod End. The terminal bag end of a shrimp trawl.

Coelenterates. An invertebrate animal of the phylum Coelenterata; i.e., corals, sea anemones, jellyfish.

Coelom. The body cavity between the body wall and intestines in many-celled animals.

Cusp. A pointed projection.

Dimorphism. The state of individuals in a population occurring in two forms; i.e., male and female.

Dinoflagellate. One of many planktonic, chiefly marine, plant-like flagellates.

Distal. The terminal portion of a limb, situated away from the point of attachment.

Diurnal. Active during daylight.

Dugong. A member of the family Trichechidae, related to the manatee, occurring in the Indian Ocean.

Elasmobranch. Pertaining to the group of cartilaginous fishes, sharks and rays.

Epibiont. An organism growing upon, or on the outside, of an object.

Fecundity. The ability to produce offspring.

Foredune. A low dune, often occupied by sand-binding beach grasses, bordering the shore.

Formalin. A solution of formaldehyde in water.

Frontal. The beach zone nearest the water.

Gamete. The reproductive cell which unites with another to form the cell that develops into a new individual.

Gastrula. The developmental stage which follows blastula.

Glottis. The opening of the throat.

Gravid. Containing eggs, as opposed to pregnant in livebearers.

Headstart. To provide a growth advantage.

Herpetologist. A person specializing in the study of amphibians and reptiles.

Hypothermia. A below normal, or less than optimum, body temperature.

Intertidal. The area of shore located between the elevations of high and low tides.

Keratinous. Composed of substance formed from horn-like material; i.e., horn, fingernails, or claws.

Laparoscopy. A surgical procedure inspection, with a lighted instrument, of organs inside the body cavity.

Leucistic. Lacking all pigment, except in the eyes which are normal.

Littoral. Pertaining to the shore of a body of water, i.e., the ocean.

Mangrove. Any of a variety of coastal trees growing in large colonies in tidal zones of the tropics; i.e., the red mangrove.

Melanism. Abnormal development of dark pigment in the skin.

Monotypic. Having only one representative.

Myopic. Nearsighted.

Necropsy. An autopsy examination performed on an animal.

Necrosis. Death of tissue.

Neonate. A newly hatched sea turtle.

Nocturnal. Active at nighttime.

Olfactory. Pertaining to the sense of smell.

Ostium. Pertaining to a small opening.

Paleontologist. A person who specializes in prehistoric forms of life by studying fossil plants and animals.

Pelagic. Occurring at sea, far from land.

Phalanges. Hand and foot bones.

Phototactic. The movement of an organism toward or away from a light source.

Pound Net. A fish trap composed of staked nets arranged to form an enclosure which has a small opening into the encircling net system.

Propagule. A part of a plant (mangroves) that when separated from the plant of origin will produce another individual.

Reptilia. The vertebrate class which includes amphisbaenids, crocodilians, lizards, snakes, turtles, and the tuatara.

Rookery. Used in sea turtle biology to indicate a dense nesting colony.

Salinity. The degree of being saline or saltiness.

Scutes. Large dermal keratinous plates.

Spat. The spawn of shellfish.

Subtidal. Submerged, not exposed at the lowest tide.

Taxonomy. The science of classifying organisms.
Titer. The strength of a solution.
Tunicates. Solitary or colonial sea chordates of the subphylum Tunicata.
Venter. Pertaining to the bottom.
Vitelline Membrane. The cell wall of an egg.

Suggested Reading

There are many books available relative to sea turtles and other groups of chelonians. I have listed below the major works which include coverage of marine turtles or contain sections devoted to them. Some of the older books may be out of print; however, there have been some excellent books published in recent years. All of the titles provided will give the reader additional information on the marine turtle group. Should some be out of print, many larger metropolitan libraries will have these books on their shelves or in their reference section. For specific information on a variety of aspects which relate to the biology or life history of these animals, you can contact your nearest museum or university library and obtain a wealth of information from the scientific literature available.

Alderton, David. 1988. *Turtles and Tortoises of the World.* Facts on File. New York.

Bustard, H.R. 1972. *Sea Turtles. Their Natural History and Conservation.* W. Collins & Sons. London.

Carr, A.F., Jr. 1952. *Handbook of Turtles.* Cornell University Press. Ithaca, New York.

Carr, A.F., Jr. 1984. *So Excellent a Fishe* (revised edition). Charles Scribner's Sons. New York.

Ernst, C.H., and R.W. Barbour. 1972. *Turtles of the United States.* University Presses of Kentucky, Lexington.

Ernst, C.H., and R.W. Barbour. 1989. *Turtles of the World.* Smithsonian Institution Press. Washington, D.C.

Hopkins, S. and J. Richardson (Editors). 1984. *Recovery Plan for Marine Turtles.* National Marine Fisheries Service. Washington, D.C.

Obst, F.J. 1988. *Turtles, Tortoises and Terrapins.* St. Martin's Press. New York.

Pope, C.H. 1939. *Turtles of the United States and Canada.* Alfred A. Knopf. New York.

Pritchard, P.C.H. 1967. *Living Turtles of the World.* T.F.H. Publications, Inc. Neptune, New Jersey.

Pritchard, P.C.H. 1979. *Encyclopedia of Turtles.* T.F.H. Publications, Inc. Neptune, New Jersey.

Rebel, T.P. 1974. *Sea Turtles and the Turtle Industry of the West Indies, Florida and the Gulf of Mexico* (revised edition). University of Miami Press. Coral Gables, Florida.

Rudloe, J. 1979. *Time of the Turtle.* Alfred A. Knopf. New York.

Index

F

G

T

U

V

W

Y

Z